The Psychology of
Intelligence

Jean
Piaget

The Psychology of
Intelligence

Translated by Malcolm Piercy
and D. E. Berlyne

 London and New York

La Psychologie de l'intelligence first published 1947
by Armand Colin, Paris

First English edition published 1950
by Routledge & Kegan Paul

First published in Routledge Classics 2001
by Routledge
11 New Fetter Lane, London EC4P 4EE
29 West 35th Street, New York, NY 10001

Routledge is an imprint of the Taylor & Francis Group

© 1950 Jean Piaget; translated from the French by Malcolm Piercy
and D. E. Berlyne

Typeset in Joanna by RefineCatch Limited, Bungay, Suffolk
Printed and bound in Great Britain by
TJ International Ltd, Padstow, Cornwall

British Library Cataloguing in Publication Data
A catalogue record for this book is available from the British Library

Library of Congress Cataloging in Publication Data
A catalog record for this book has been applied for

ISBN 0–415–25401–9

PREFACE

A book on the "Psychology of Intelligence" could cover half the realm of psychology. The following pages are confined to outlining one view, that based on the formation of "operations," and to determining as objectively as possible its place among others which have been put forward. The first task is to define intelligence in relation to adaptive processes in general (Chap. 1), then to show, by examining the "psychology of thought", that the act of intelligence consists essentially in "grouping" operations according to certain definite structures (Chap. 2). Then, if intelligence is thus conceived as the form of equilibrium towards which all cognitive processes tend, there arises the problem of its relations with perception (Chap. 3), and with habit (Chap. 4); as well as the question of its development (Chap. 5) and of its socialization (Chap. 6).

In spite of the abundance and the value of well-known studies, the psychological theory of intellectual mechanisms is only in its infancy, and we are barely beginning to glimpse the sort of precision of which it might be capable. It is this feeling of research in progress that I have sought to express.

This little volume contains the substance of the lectures that I had the privilege of giving at the Collège de France in 1942 at an hour when university men felt the need to show their solidarity in the face of violence and their fidelity to permanent values. It is difficult for me, as I rewrite these pages, to forget the welcome given by my audience, as well as the contact which I had at that time with my friends.

J.P.

PREFACE TO THE SECOND (FRENCH) EDITION

The reception given to this little work has in general been a favourable one, which gives us the courage to reprint it without any alterations. Nevertheless, one criticism has frequently been levelled at our conception of intelligence—that it makes no reference to the nervous system or to its maturation in the course of the individual's development. That, we think, is a simple misunderstanding. Both the concept of "assimilation" and the transition from rhythms to regulations and from these to reversible operations demand a neurological as well as a psychological (and logical) interpretation. And these two interpretations, far from contradicting each other, can only agree. We shall explain ourselves elsewhere on this essential point, but we have never felt entitled to deal with it before completing the detailed psychogenetic researches which are summed up in this little book.

NOTE

The translators desire to thank
Messrs. P. F. C. Castle and C. Gattegno
for many valuable suggestions.

CONTENTS

"groupings". Classification of "groupings" and of the fundamental operations of thought. Equilibrium and development.

Part I

The Nature of Intelligence

1

INTELLIGENCE AND BIOLOGICAL ADAPTATION

Every psychological explanation comes sooner or later to lean either on biology or on logic (or on sociology, but this in turn leads to the same alternatives). For some writers mental phenomena become intelligible only when related to the organism. This view is of course inescapable when we study the elementary functions (perception, motor functions, etc.) in which intelligence originates. But we can hardly see neurology explaining why 2 and 2 make 4, or why the laws of deduction are forced on the mind of necessity. Thus arises the second tendency, which consists in regarding logical and mathematical relations as irreducible, and in making an analysis of the higher intellectual functions depend on an analysis of them. But it is questionable whether logic, regarded as something eluding the attempts of experimental psychology to explain it, can in its turn legitimately explain anything in psychological experience. Formal logic, or logistics, is simply the axiomatics of states of equilibrium of thought, and the positive science corresponding central role of intelligence in mental life and in the life of the organism itself; intelligence, the most plastic and at the same time the most durable structural equilibrium of behaviour, is essentially a system of living and acting operations. It is the most highly developed form of mental adaptation, that is to say, the indispensable instrument for interaction between the subject and the universe when the scope of this interaction goes beyond immediate and momentary contacts to achieve far-reaching and stable relations. But, on the other hand, this use of the term precludes our determining where intelligence starts; it is an ultimate goal, and its origins are indistinguishable from those of sensori-motor adaptation in general or even from those of biological adaptation itself.

ADAPTIVE NATURE OF INTELLIGENCE

If intelligence is adaptation, it is desirable before anything else to define the latter. Now, to avoid the difficulties of teleological language, adaptation must be described as an equilibrium between the action of the organism on the environment and vice versa. Taking the term in its broadest sense, "assimilation" may be used to describe the action of the organism on surrounding objects, in so far as this action depends on previous behaviour involving the same or similar objects. In fact every relation between a living being and its environment has this particular characteristic: the former, instead of submitting passively to the latter, modifies it by imposing on it a certain structure of its own. It is in this way that, physiologically, the organism absorbs sub-stances and changes them into something compatible with its own substance. Now, psychologically, the same is true, except that the modifications with which it is then concerned are no longer of a physico-chemical order, but entirely functional, and are determined by movement, perception or the interplay of real or potential actions (conceptual operations, etc.). Mental assimilation is thus the incorporation of objects into patterns of behaviour, these patterns being none other than the whole gamut of actions capable of active repetition.

Conversely, the environment acts on the organism and, following the practice of biologists, we can describe this converse action by the term "accommodation", it being understood that the individual never suffers the impact of surrounding stimuli as such, but they simply modify the assimilatory cycle by accommodating him to themselves. Psychologically, we again find the same process in the sense that the pressure of circumstances always leads, not to a passive submission to them, but to a simple modification of the action affecting them. This being so, we can then define adaptation as an equilibrium between assimilation and accommodation, which amounts to the same as an equilibrium of interaction between subject and object.

Now in the case of organic adaptation, this interaction, being of a material nature, involves an interpenetration between some part of the living body and some sector of the external environment. Psychological life, on the other hand, begins, as we have seen, with functional interaction, that is to say, from the point at which assimilation no longer alters assimilated objects in a physico-chemical manner but simply incorporates them in its own forms of activity (and when accommodation only modifies this activity). We can then understand that, superimposed on the direct interpenetration of organism and environment, mental life brings with it indirect interaction between subject and object, which takes effect at ever increasing spatio-temporal distances and along ever more complex paths. The whole development of mental activity from perception and habit to symbolic behaviour and memory, and to the higher operations of reasoning and formal thought, is thus a function of this gradually increasing distance of interaction, and hence of the equilibrium between an assimilation of realities further and further removed from the action itself and an accommodation of the latter to the former.

It is in this sense that intelligence, whose logical operations constitute a mobile and at the same time permanent equilibrium between the universe and thought, is an extension and a perfection of all adaptive processes. Organic adaptation, in fact, only ensures an immediate and consequently limited equilibrium between the individual and the present environment. Elementary cognitive functions, such as perception, habit and memory, extend it in the direction of present space (perceptual contact with distant objects) and of short-range reconstructions and anticipations. Only intelligence, capable of all its detours and reversals by action and by thought, tends towards an all-embracing equilibrium by aiming at the assimilation of the whole of reality and the accommodation to it of action, which it thereby frees from its dependence on the initial *hic* and *nunc*.

DEFINITION OF INTELLIGENCE

If we undertake to define intelligence, which is certainly important for determining the field which we shall be studying under this heading, it is sufficient that we be agreed on the degree of complexity of distant interaction which we shall call "intelligent". But here difficulties arise, since the lower demarcation line remains arbitrary. For some, such as Claparède and Stern, intelligence is a mental adaptation to new circumstances. Thus Claparède opposes intelligence to instinct and habit, which are hereditary or acquired adaptations to recurring circumstances ; but for him it begins with the most elementary empirical trial-and-error (the origin of the implicit trial-and-error which subsequently characterizes the search for a hypothesis). For K. Bühler, who also divides mental structures into three types (instinct, training and intelligence), this definition is too broad; intelligence only appears with acts of insight (*Aha-Erlebnis*), while trial-and-error is a form of training. Köhler likewise reserves the term intelligence for acts of abrupt restructuring and excludes trial-and-error. It cannot be denied that the latter appears right from the formation of the simplest habits, which are themselves, when they are first formed, adaptations to new circumstances. On the other hand, problem, hypothesis, and control, whose combination is the mark of intelligence according to Claparède also, already exist in embryo in the needs, the trials-and-errors and the empirical test characteristic of the least developed sensori-motor adaptations. We must therefore choose between these two alternatives: either we must be satisfied with a functional definition at the risk of encompassing almost the entire range of cognitive structures, or else we must choose a particular structure as our criterion, but the choice remains arbitrary and runs the risk of overlooking the continuity which exists in reality.

However, it is still possible to define intelligence by the direction towards which its development is turned, without insisting on the question of boundaries, which become a matter of stages or of successive forms of equilibrium. We can therefore regard the matter from the point of view both of the functional situation and of the structural mechanism. From the first of these points of view, we can say that behaviour becomes more "intelligent" as the pathways between the subject and the objects on which it acts cease to be simple and become progressively more complex. Thus perception only requires simple paths, even if the object perceived is very remote. A habit might seem more complex, but its spatio-temporal articulations are welded into a unique whole with no independent or separable parts. An act of intelligence, on the other hand, such as finding a hidden object or recognizing the meaning of a picture, involves a certain number of paths (in space and time) which can be both isolated and synthesized. Thus, from the point of view of the structural mechanism, elementary sensori-motor adaptations are both rigid and unidirectional, while intelligence tends towards reversible mobility. That, as we shall see, is the essential property of the operations which characterize living logic in action. But we can see straight away that reversibility is the very criterion of equilibrium (as physicists have taught us). To define intelligence in terms of the progressive reversibility of the mobile structures which it forms is therefore to repeat, in different words, that intelligence constitutes the state of equilibrium towards which tend all the successive adaptations of a sensori-motor and cognitive nature, as well as all assimilatory and accommodatory interactions between the organism and the environment.

CLASSIFICATION OF POSSIBLE INTERPRETATIONS OF INTELLIGENCE

From the biological point of view, intelligence thus appears as one of the activities of the organism, while the objects to which it adapts itself constitute a particular sector of the surrounding environment. But as the knowledge that intelligence builds up achieves a privileged equilibrium, because this is the necessary limit of sensori-motor and symbolic interaction, while distances in space and time become indefinitely extended, intelligence engenders scientific thought itself, including biological knowledge. It is therefore natural that the psychological theories of intelligence should come to be placed among biological theories of adaptation and theories of knowledge in general. It is not surprising that there should be some relationship between psychological theories and epistemological doctrines since, even if psychology has been freed from philosophical tutelage, there happily remains some bond between the study of mental functions and that of the processes of scientific knowledge. But what is more interesting is that there exists a parallelism, and a fairly close one, between the great biological doctrines of evolutionary variation (and therefore of adaptation) and the particular theories of intelligence as a psychological fact; psychologists have, in fact, often been unaware of the currents of biological inspiration behind their interpretations, just as biologists have sometimes unwittingly adopted one particular psychological position among other possible ones (Cf. the role of habit in Lamarck or of competition and strife in Darwin); moreover, in view of the affinity between the problems, there may be a simple convergence of solutions and so the latter may confirm the former.

From the biological point of view, the relations between the organism and the environment admit of six possible interpretations according to the following combinations (each of which has led to its own solution, classical or contemporary): either (I) we reject the idea of a genuine evolution, or else (II) we admit its existence; then, in both cases (I and II) we attribute adaptations (1) to factors external to the organism, or (2) to internal factors, or (3) to an interaction between the two. So (I) from the non-evolutionist point of view, we may attribute adaptation (I1) to a pre-established harmony between the organism and the properties of the environment, (I2) to a preformism allowing the organism to respond to every situation by actualizing its potential structures, or else (I3) to the "emergence" of complete structures, irreducible to elements and determined simultaneously from within and from without.[1]

---

[1] Pre-established harmony (I1) is the solution inherent in classical creationism and it constitutes the only explanation of adaptation which is in fact at the disposal of vitalism in its pure form. Preformism (I2) has sometimes been associated with vitalist solutions, but it can become independent of them and often persists in mutationist guises among authors who deny all constructive character to evolution and consider every new characteristic as the actualization of potentialities which hitherto were merely latent. Conversely, the view based on emergence (I3) reverts to explaining the innovations which arise in the hierarchy of beings by complex structures which are irreducible to the elements of the previous level. From these elements there "emerges" a new totality, which is adaptive because it unites in an indissociable whole both the

As for the evolutionist points of view (II), they likewise explain adaptive variations, by environmental pressure (Lamarckism II1), or by endogenous mutations with subsequent selection (mutationism II2)[1], or (II3) by a progressive inter-action between internal and external factors.

Now it is striking to note how we find the same broad currents of thought in the interpretation of knowledge itself, regarded as a relationship between the thinking subject and objects. Corresponding to the pre-established harmony of cre-ationist vitalism, there is (I1) the realism of those doctrines which see in reason an innate adaptation to eternal forms or essences; corresponding to preformism, there is (I2) apriorism which explains consciousness by internal structures which precede experience; and corresponding to the "emergence" of new structures there is (I3) contemporary phenomenology, which simply analyses the various forms of thought, refusing either to derive them genetically from each other or to distinguish in them the roles of subject and object. Evolutionist interpretations, on the other hand, reappear in those epistemological schools which allow for the progressive development of reason; corresponding to Lamarck-ism there is (II1) empiricism, which explains knowledge by the pressure of objects; corresponding to mutationism there are (II2) conventionalism and pragmatism, which attribute the fittingness of mind to reality to the untrammelled creation of subjective ideas, subsequently selected according to a principle of simple expediency. Finally, interactionism (II3) involves a

internal mechanisms and their relations with the external environment. While admitting the fact of evolution, the hypothesis of emergence thus reduces it to a series of syntheses, each irreducible to the others, so that it is broken up into a series of distinct creations.

[1] In mutationist explanations of evolution subsequent selection is due to the environment itself. In Darwin it was attributed to competition.

relativism, which would describe knowledge as the product of an indissociable collaboration between experience and deduction.

Without insisting on this parallelism in its most general form, we may now note how contemporary strictly psychological theories of intelligence are inspired by the same currents of thought, whether biological emphasis is dominant or whether philosophical influences related to the study of knowledge are felt.

There is no doubt, to begin with, that a fundamental incompatibility divides two kinds of interpretations: those which, while recognizing the existence of the facts of development, cannot help considering intelligence as a primary datum, and thus reduce mental evolution to a sort of gradual awakening of consciousness without any real construction of anything, and those which seek to explain intelligence by its own development. It should be noted moreover that the two schools collaborate in the discovery and analysis of actual experimental facts. That is why it is fitting to classify objectively all contemporary all-embracing interpretations, inasmuch as they have helped to throw light on one particular aspect or another of the facts to be explained; the demarcation line between psychological theories and philosophical doctrines is in fact to be found in this appeal to experience, and not in the initial hypotheses.

Among the non-evolutionist theories, there are first of all (I1) those which remain constantly faithful to the idea of an intelligence-faculty, a sort of direct knowledge of physical entities and of logical or mathematical ideas by a pre-established harmony between intellect and reality. We must confess that few experimental psychologists still adhere to this hypothesis. But the problems arising from the common frontiers of psychology and the analysis of mathematical thought have caused certain symbolic logicians, e.g. Bertrand Russell, to formulate such a conception of intelligence and even to wish to impose it on psychology itself (cf. his *Analysis of Mind*).[1]

A more prevalent hypothesis (I2) is that according to which intelligence is determined by internal structures, which are likewise not formed but gradually become explicit in the course of development, owing to a reflection of thought on itself. This apriorist current has in fact inspired a good deal of the work of the German *Denkpsychologie* and is consequently found at the root of numerous experimental researches on thought, using the familiar methods of introspection, which have been developing from 1900–1905 to the present day. Naturally this does not mean that every use of these methods of investigation leads to this explanation of intelligence: Binet's work testifies to the contrary. But for K. Bühler, Selz and many others, intelligence eventually became, as it were, "a mirror of logic", which imposes itself from within with no possible causal explanation.

In the third place (I3), corresponding to emergence and phenomenology (with the actual historical influence of the latter), there is a recent theory of intelligence which has raised the problem anew in a very suggestive way: the Configuration (*Gestalt*) theory. The notion of a "complex configuration", resulting from experimental researches in perception, involves the assertion that a whole is irreducible to the elements which compose it, being governed by special laws of organization or equilibrium. Now, having analysed these laws of structuring in the realm of perception and having come across them again in motor functions, memory, etc., the Configuration theory has been applied to intelligence itself, both in its reflective (logical thought) and its sensori-motor form (intelligence in animals

[1] The author desires to indicate that his discussion of Russell's views on this and subsequent pages refers only to that writer's first period. Russell has since rejected this position in favour of an extreme empiricism. (*Translator's note.*)

and in children at the pre-linguistic stage). Thus Köhler, in connection with chimpanzees, and Wertheimer, in connection with
the syllogism, etc., have spoken of "immediate restructurings"
seeking to explain the act of insight by the "goodness" (*Prägnanz*)
of well organized structures, which are neither endogenous nor
exogenous but embrace subject and object in a total field. Furthermore, these *Gestalten*, which are common to perception,
movement and intelligence, do not evolve, but represent permanent forms of equilibrium, independent of mental development (we may in this respect find all intermediate stages
between apriorism and the Configuration theory, although the
latter is normally found linked with a physical or physiological
realism of "structures").

Such are the three principal non-genetic theories of intelligence. It may be noted that the first reduces cognitive adaptation
to pure accommodation, since it sees thought only as the mirror
of ready-made "ideas", that the second reduces it to pure assimilation, since it regards intellectual structures as exclusively
endogenous, and that the third unites assimilation and accommodation in a single whole, since, from the *Gestalt* point of view,
there exists only the field linking objects and the subject, with
neither activity on his part nor the isolated existence of the
object.

As for genetic interpretations, we find once more those which
explain intelligence in terms of the external environment only
(associationist empiricism corresponding to Lamarckism), the
activity of the subject (the trial-and-error theory at the level of
individual adaptation, corresponding to mutationism at the level
of hereditary variations) and the relationship between subject
and object (operational theory).

Empiricism (II1) is scarcely upheld any longer in its pure
associationist form, except for some authors, of predominantly
physiological interests, who think they can reduce intelligence
to a system of "conditioned" responses. But we find less rigid

forms of empiricism in Rignano's interpretations, which reduce reasoning to mental experience, and especially in Spearman's interesting theory, which is both statistical (factor analysis of intelligence) and descriptive; from this second point of view, Spearman reduces the operations of intelligence to the "apprehension of experience" and to the "eduction" of relations and "correlates", that is to say, to a more or less complex reading of immediately given relations. These relations, then, are not constructed but discovered by simple accommodation to external reality.

The notion of trial-and-error (II2) has given rise to several interpretations of learning and of intelligence itself. The trial-and-error theory elaborated by Claparède constitutes in this respect the most far-reaching exposition: intelligent adaptation consists of trials or hypotheses, due to the activity of the subject, and of their selection, effected afterwards under the pressure of experience (successes or failures). This empirical control, which from the outset selects the subject's trials, is subsequently internalized in the form of anticipations due to awareness of relations, just as motor trial-and-error is extended into symbolic trial-and-error or imagination of hypotheses.

Finally, emphasizing the interaction of the organism and the environment leads to the operational theory of intelligence (II3). According to this point of view, intellectual operations, whose highest form is found in logic and mathematics, constitute genuine actions, being at the same time something produced by the subject and a possible experiment on reality. The problem is therefore to understand how operations arise out of material action, and what laws of equilibrium govern their evolution; operations are thus concerned as grouping themselves of necessity into complex systems, comparable to the "configurations" of the Gestalt theory, but these, far from being static and given from the start, are mobile and reversible, and round themselves off only when the limit of the individual and social genetic process that characterizes them is reached.[1]

This sixth point of view is the one we shall develop. As for trial-and-error theories and empiricist conceptions, we shall discuss them with particular reference to sensori-motor intelligence and its relations with habit (Chap. 4). The Configuration theory necessitates a special discussion, which we shall focus upon the important problem of the relations between perception and intelligence (Chap. 4). As for the doctrine of an intelligence pre-adapted to independently subsisting logical entities and that of a thought reflecting an *a priori* logic, we shall return to them at the beginning of the next chapter. In fact these both raise what we may call the "preliminary question" of the psychological study of intellect: may we hope for a real explanation of intelligence, or does intelligence constitute a primary irreducible fact, being the mirror of a reality prior to all experience, namely logic?

---

[1] We should note in this respect that, although the social nature of operations follows from their character as effective action and their gradual grouping, we shall nevertheless, for the sake of clarity of exposition reserve the discussion of Social factors in thought until Chapter VI.

2

"THOUGHT PSYCHOLOGY" AND THE PSYCHOLOGICAL NATURE OF LOGICAL OPERATIONS

How far a psychological explanation of intelligence is possible depends on the way in which logical operations are interpreted: are they the reflection of an already formed reality or the expression of a genuine activity? No doubt only the notion of an axiomatic logic can enable us to escape from this dilemma, by submitting the actual operations of thought to a genetic interpretation, while admitting the irreducible character of their formal connections when these are analysed axiomatically; the logician then proceeds as does the geometer with the space that he constructs deductively, while the psychologist can be likened to the physicist, who measures space in the real world. In other words, the psychologist studies the way in which the actual equilibrium of actions and operations is constituted, while the logician analyses the same equilibrium in its ideal form, i.e. as it would be if it were completely realised, and as it is imposed on the mind as a norm.

BERTRAND RUSSELL'S INTERPRETATION

We shall start from Bertrand Russell's theory of intelligence, which is marked by the maximum possible subordination of psychology to logistics. According to Russell, when we perceive a white rose we conceive at the same time the ideas of the rose and of whiteness, and this by a process analogous to that of perception; we apprehend directly, and as if from without, the "universals" corresponding to perceptible objects and "subsisting" independently of the subject's thought. But what then of false ideas? These are ideas as much as any others, and the qualities of false and true are applied to concepts just as there are red roses and white roses. As for the laws which govern universals and which control their relations, they depend on logic alone, and psychology can only bow before this previous knowledge which is given to it ready made.

This is the hypothesis. It is no use accusing it of being metaphysical or metapsychological just because it runs counter to the common sense of experimentalists; the mathematician's common sense finds it quite acceptable and psychology must take mathematicians into account. So radical a thesis is even well worth pondering over. First of all, it does away with the notion of an operation, since, if we apprehend universals from without, we do not construct them. In the expression $1 + 1 = 2$, the sign $+$ signifies nothing more than a relation between the two unities and in no way an activity producing the number 2; as Couturat has clearly indicated, the notion of an operation is essentially "anthropomorphic". Russell's theory therefore dissociates *a fortori* the subjective factors of thought (belief, etc.) from the objective factors (necessity, probability, etc.). In fact it rejects the genetic point of view; an English follower of Russell once said, in order to prove the uselessness of research on thought in children, that "the logician is interested in true ideas, while the psychologist finds pleasure in describing false ones."

But, if we have seen fit to begin this chapter with a review of Russell's ideas, it was in order that we might note at once that the demarcation line between the knowledge derived from symbolic logic and psychology cannot be crossed by the former with impunity. Even if, from the axiomatic point of view, the operation were to appear devoid of significance, its very "anthropomorphism" would make a mental reality of it. From the genetic point of view, operations are indeed genuine actions and do not consist merely of taking note of or apprehending relations. When 1 is added to 1 what happens is that the subject combines two units into one whole, when he could keep them apart. There is no doubt that this action, occurring in thought, acquires a character *sui generis* which distinguishes it from other actions; it is reversible, i.e., having combined the two units, the subject can then separate them and thus find himself where he started. But this does not make it any the less a genuine action, radically different from the simple reading of a relation such as 2>1. Now to this followers of Russell will only reply with a non-psychological argument: it is an illusory action, since 1 + 1 have made 2 from all eternity (or, as Carnap and Wittgenstein would say, since 1 + 1 = 2 is only a tautology, characteristic of the language of "logical syntax", and does not concern thought itself, whose functioning is specifically experimental). Broadly speaking, mathematical thought is mistaken when it believes it can construct or invent, since it is confined to revealing the various aspects of an already formed world (and, according to the Vienna circle, an entirely tautological one). However, if we deny the psychology of intelligence the right to concern itself with the nature of logico-mathematical entities, the fact remains that individual thought cannot remain passive in the face of ideas (or of the symbols of a logical language) any more than it can in the presence of physical entities, and that in order to assimilate them it has to reconstruct them by means of psychologically real operations.

We may add that the assertions of Bertrand Russell and the Vienna circle, regarding the independent existence of logico-mathematical entities and the operations which seem to engender them, are just as arbitrary from the purely logical point of view as they are from the psychological: in fact they will always meet the fundamental difficulty inherent in a realism of classes, relations and numbers, namely, that of the antinomies relating to the "class of all classes" and to infinity. On the other hand, from the operational point of view, infinite entities are only the expression of operations capable of being repeated indefinitely.

Finally, from a genetic point of view, the hypothesis of a direct apprehension by thought of universals, subsisting independently of it, is even more chimerical. We may admit that the false ideas of the adult have an existence comparable to that of true ideas. What then are we to think of the concepts successively constructed by the child in the course of the different stages of his development? Do the "schemata" of preverbal practical intelligence "subsist" outside the subject? And what of those of animal intelligence? If we reserve eternal "subsistence" solely for true ideas, at what age does their apprehension begin? And, furthermore, even if stages of development simply mark successive approximations of intelligence in its conquest of immutable "ideas", what proof have we that the normal adult or the logicians of Russell's school have succeeded in grasping them and will not be continually surpassed by future generations?

"THOUGHT PSYCHOLOGY": K. BÜHLER AND SELZ

The difficulties we have just encountered in Russell's interpretation of intelligence recur in part in the interpretation arrived at by the German *Denkpsychologie*, although in this case it is the work of pure psychologists. It is true that for the writers of this school logic is not imposed on the mind from without but from within; the conflict between the exigencies of psychological explanation and those of the logicians' deduction is certainly attenuated by it; but, as we shall see, it is not entirely assuaged, and the shadow of formal logic continues as an irreducible datum to dog the explanatory and causal research of the psychologist as long as he does not adopt a thoroughgoing genetic point of view. Now the German "thought psychologists" have in fact been inspired either by essentially apriorist trends or by phenomenological trends (the influence of Husserl has been particularly clear) with all intermediate stages between the two.

As a method, the psychology of thought came into being simultaneously in France and in Germany. Turning away entirely from the associationism which he defended in his little book, *La Psychologie du raisonnement*, Binet reconsidered the question of the relations between thought and images by an interesting method of controlled introspection, and by this means he discovered the existence of imageless thought; in 1903, in his *Etude expérimentale de l'intelligence*, he maintains that relations, judgments, attitudes, etc. go beyond imagery, and thinking cannot be reduced to "looking at pictures." As for knowing what these acts of thought which resist an associationist interpretation consist of, Binet reserves his opinion, confining himself to noting the relationship between intellectual and motor "attitudes", and concludes that, from the point of view of introspection alone, "thought is an unconscious activity of the mind." This is extremely instructive but certainly a disappointing test of the resources of a method which is thus shown to be more fruitful in raising problems than in solving them.

In 1900, Marbe (*Experimentelle Untersuchungen über das Urtheil*) also enquired how judgment differed from association and likewise hoped to resolve the question by a method of controlled introspection. Marbe meets with a most varied range of states of consciousness: verbal representations, images, sensations of movement, attitudes (doubt, etc.), but nothing constant. Although he notes that the necessary condition for judgment is the voluntary or intentional character of the report, he does not consider this condition as sufficient, and concludes with a denial which recalls Binet's formula: there is no state of consciousness which is invariably associated with judgment and which can be regarded as its determinant. But he adds, and this to us seems to have influenced directly or indirectly all German *Denkpsychologie*, that judgment consequently implies the intervention of a factor that is non-psychological because it comes from pure logic. We see that we were not exaggerating when we forecast the reappearance, on this new plane, of the difficulties inherent in the logicalism of the Platonists.

Next came the work of Watt, Messer and K. Bühler, inspired by Külpe, for which the Würzburg school is famous. Watt, using the method of controlled introspection, studies the associations reported by the subject following instructions (e.g. supraordinate associations, etc.) and finds that the task may act together with images, or in an imageless state of consciousness (*Bewusstheit*), or even unconsciously. He therefore formulates the hypothesis that Marbe's "intention" is just the effect of the task (whether external or internal), and thinks that he can solve the problem of judgment by showing it to be a series of states conditioned by a mental factor which was at one time conscious and still exerts its influence.

Messer finds Watt's description too vague, since it is applied to a controlled response as well as to judgment, and he takes up the problem again with a similar technique: he distinguishes between constrained association and judgment, which is something either accepted or rejected, and devotes the main body of his work to analysing the different mental types of judgment.

Finally, with K. Bühler we reach the culmination of the work of the Würzburg school. The poverty of the initial results produced by the method of controlled introspection seems to him to result from the fact that the questions used involved processes which were too simple, and thenceforward he undertakes to analyse with his subjects the solution of genuine problems. The elements of thought obtained by this procedure fall into three categories: images whose role is accessory, and not essential as associationism would have it; intellectual feelings and attitudes; and, above all, "thoughts" themselves (*Bewusstheiten*). These for their part occur in the form of "consciousness of relation" (e.g. A < B), "consciousness of rules" (e.g. thinking of the inverse square of the distance without knowing what objects or what distances are involved), or of "purely formal intentions" in the scholastic sense (e.g. thinking of the architecture of a system). Thus conceived, the psychology of thought arrives at a precise and often very refined description of intellectual states, but one that is analogous to logical analysis and in no way explains operations as such.

The work of Selz, on the other hand, goes beyond the results of the Würzburg school towards an analysis of the actual dynamics of thought and not merely of its isolated states. Selz, like Bühler, studies the solution of actual problems, but he attempts not so much to describe the elements of thought as to understand how the solutions are reached. Having studied "reproductive thought" in 1913, he tries, in 1922, (*Zur Psychologie des produktiven Denkens und des Irrtums*) to penetrate the secret of mental construction. It is interesting to note that the more research is directed towards the actual activity of thought, the further it departs from logical atomism, which consists in classing relations, judgments and isolated schemata, and the nearer it comes to living wholes after the pattern of Gestalt Psychology, there being a different pattern where operations are concerned, as we shall shortly find. In fact for Selz all thinking activity consists of completing a whole (theory of *Komplexergänzung*): the solution of a problem cannot be reduced to the stimulus-response schema, but consists of filling in the gaps in "complexes" of ideas and relations. When a problem is put, two possibilities thus present themselves. It may be only a question of reconstruction, no new construction being required, and the solution consists simply in having recourse to already existing "complexes" there is then "actualization of knowledge" and therefore thought which is simply "reproductive". Or else it may be a genuine problem, testifying to the existence of gaps within the complexes hitherto adopted, and so it is no longer a matter of utilizing knowledge but of utilizing methods of solution (applying known methods to a new case), or even of deriving new methods from old ones; there is, in these last two cases, "productive" thought, and this is really where completing wholes or already existing complexes comes in. As for this "filling in of gaps", it is always orientated by "anticipatory schemata" (comparable to Bergson's "dynamic schema"), which weave between new data and the main body of the corresponding complex a system of provisional global relations constituting the pattern of the solution to be found (i.e. the directing hypothesis). These relations themselves are finally made more precise by a mechanism obeying exact laws; these laws are none other than those of logic, of which, when all is said and done, thought is the mirror.

We should also note Lindworsky's work, which comes between the two studies by Selz and anticipates his conclusions. As for Claparède's study of the genesis of the hypothesis, this will be discussed in relation to trial-and-error (Chapter 4).

CRITIQUE OF "THOUGHT PSYCHOLOGY"

It is clear that the researches just mentioned have rendered considerable service to the study of intelligence. They have freed thought from the conception of the image as a constituent element, and have discovered, like Descartes, that judgment is an act.

They have accurately described the various states of thought and have thus shown, contradicting Wundt, that introspection may be "controlled", i.e. systematized by an observer.

But first we should mention that even at the level of simple description the relations between image and thought have been over-simplified by the Würzburg school. It remains an acknowledged fact that the image does not constitute an element of thought itself. It merely accompanies it and serves as a symbol for it, an individual symbol completing the collective signs of language. The "Meaning" school, inspired by Bradley's logic, has clearly shown that all thought is a system of meanings, and it is this notion that Delacroix and his pupils, in particular Meyerson, have developed in connection with the relations between thought and the image. These meanings involve in fact "significates", which constitute thought itself, but also "significants", comprising verbal signs or imaged symbols which are formed hand in hand with thought.

On the other hand, it is obvious that the very method used by *Denkpsychologie* prevents it from going beyond pure description, and that it fails to explain the actual constructive mechanisms of intelligence, because introspection, even when controlled, surely deals only with the products of thought and not with its formation. Furthermore, it is restricted to subjects capable of reflection; whereas we should perhaps look for the secret of intelligence in children under the age of seven or eight!

Lacking in this way any genetic background, "Thought Psychology" analyses only the final stages of intellectual development. Speaking in terms of states and of completed equilibrium, it is not surprising that it arrives at a panlogicism and is obliged to give up psychological analysis when faced with the irreducibly given nature of the laws of logic. From Marbe, who invoked logical law directly as a non-psychological factor which intervenes causally and fill the gaps of mental causality, to Selz who arrived at a sort of logico-psychological parallelism by making thought the mirror of logic, logical fact remains for all these writers inexplicable in psychological terms.

No doubt Selz freed himself partially from the unduly narrow procedure of analysing states and elements in order to try to follow the dynamics of the act of intelligence. So he discovers the wholes which characterize systems of thought, as well as the role of anticipatory schemata in the solution of problems. But while he frequently notes the analogies between these processes and organic and motor mechanisms, he does not trace their genetic formation. So he also joins the Würzburg school in their panlogicism, and even does so in a paradoxical manner, an example which merits reflection from those who wish to free psychology from the toils of logistic apriorism while seeking to explain logical fact.

Indeed, in revealing the essential role played by wholes in the functioning of thought, Selz might have drawn the conclusion that classical logic is incapable of describing reasoning in action, as it appears and takes form in "productive thought". Classical logic, even when rendered infinitely more flexible by the subtle and precise technique of the logistic calculus, remains atomistic; classes, relations and propositions are therein analysed with respect to their elementary operations (logical addition and multiplication, implications and contradictions, etc.). In order to interpret the action of anticipatory schemata and of *Komplexer-gänzung*, and thus of intellectual wholes which intervene in living and active thought, Selz would, on the contrary, have required a logic of wholes, and so the problem of the relations between intelligence, as a psychological fact, and logic itself would have been put in new terms calling for an essentially genetic solution. But Selz, having too much respect for *a priori* logical formulations despite their discontinuous and atomistic character, naturally meets them once more as the residue remaining after psychological analysis has done all it can and finds himself invoking them to explain the details of mental elaboration.

In short, "Thought Psychology" finished by making thought the mirror of logic, and in this lies the root of the difficulties it has found insurmountable. The question is then to ascertain whether it would not be better simply to reverse the terms and make logic the mirror of thought, which would restore to the latter its constructive independence.

LOGIC AND PSYCHOLOGY

Logic is the mirror of thought, and not vice versa; in *Classes, relations et nombres: essai sur les groupements de la logistique et la réversibilité de la pensée*, 1942, we were led to this point of view by the study of the formation of operations in the child, and that after having been persuaded from the outset of the justice of the postulate of irreducibility which inspires the "Thought Psychologists". This amounts to saying that logic is the axiomatics of reason, the psychology of intelligence being the corresponding experimental science. It seems to us essential to insist somewhat on this methodological point.

An axiomatics is an exclusively hypothetico-deductive science, i.e., it reduces to a minimum appeals to experience (it even aims to eliminate them entirely) in order freely to reconstruct its object by means of undemonstrable propositions (axioms), which are to be combined as rigorously as possible and in every possible way. In this way geometry has made great progress, seeking to liberate itself from all intuition and constructing the most diverse spaces simply by defining the primary elements to be admitted by hypothesis and the operations to which they are subject. The axiomatic method is thus the mathematical method *par excellence* and it has had numerous applications, not only in pure mathematics, but in various fields of applied mathematics (from theoretical physics to mathematical economics). The usefulness of an axiomatics, in fact, goes beyond that of demonstration (although in this field it constitutes the only rigorous method); in the face of complex realities, resisting exhaustive analysis, it permits us to construct simplified models of reality and thus provides the study of the latter with irreplaceable dissecting instruments. To sum up, an axiomatics constitutes a "pattern" for reality, as F. Gonseth has clearly shown, and, since all abstraction leads to a schematization, the axiomatic method in the long run extends the scope of intelligence itself.

But precisely because of its "schematic" character, an axiomatics cannot claim to be the basis of, and still less to replace, its corresponding experimental science, i.e. the science relating to that sector of reality for which the axiomatics forms the pattern. Thus, axiomatic geometry is incapable of teaching us what the space of the real world is like (and "pure economics" in no way exhausts the complexity of concrete economic facts). No axiomatics could replace the inductive science which corresponds to it, for the essential reason that its own purity is merely a limit which is never completely attained. As Gonseth also says, there always remains an intuitive residue in the most purified pattern (just as there is already an element of schematization in all intuition). This reason alone is enough to show why an axiomatics will never be the basis of an experimental science and why there is an experimental science corresponding to every axiomatics (and, no doubt, vice versa).

Thus the problem of the relations between formal logic and the psychology of intelligence is to find a solution comparable to that which has settled, after centuries of discussion, the conflict between deductive geometry and positive or physical geometry. As in the case of these disciplines, logic and the psychology of thought began by being confused and not differentiated at all; Aristotle no doubt thought he was writing a natural history of the mind (as well as of physical reality itself) by stating the laws of the syllogism. When psychology was set up as an independent science, psychologists came to understand (taking a considerable time over it) that the reflections contained in text-books of logic on the concept, judgment and reasoning did not exempt them from seeking to sort out the causal mechanism of intelligence. But as a residual effect of their original failure to draw a distinction, they still continued to think of logic as a science of reality, placed, in spite of its normative character, on the same plane as psychology, but concerned exclusively with "true thought" is opposed to thought in general, freed from all norms. Hence the deluded outlook of *Denkpsychologie*, according to which thought, a psychological fact, constitutes a reflection of logical laws. But, on the other hand, if logic were found to be an axiomatics, the pseudo-problem of these mutual relations would disappear through the interchange of status.

Now it seems obvious that the more logic repudiates the vagueness of verbal language in order to establish, under the name of symbolic logic or logistics, an algorithm with a rigour equalling that of mathematical language, the more it turns into an axiomatic technique. We know, moreover, the extent to which this technique has rapidly been linked up with the most general fields of mathematics, till symbolic logic has today acquired a scientific value independent of the particular philosophies of individual logicians (Russell's Platonism or the nominalism of the Vienna Circle). The very fact that philosophical interpret-ations leave its internal technique unchanged shows that the latter has reached the axiomatic level; symbolic logic thus con-stitutes, if for no other reason, an ideal "model" of thought.

But this being so, the relations between logic and psychology are made so much the simpler. Symbolic logic need not have recourse to psychology, since a question of fact in no way affects a hypothetico-deductive theory. Conversely, it would be absurd to invoke symbolic logic to settle an experimental question such as that of the actual mechanism of intelligence. Nevertheless, in so far as psychology undertakes to analyse the final states of equilibrium of thought, there is not a parallelism but a correspondence between this experimental knowledge and symbolic logic, just as there is a correspondence between a pattern and the reality which it represents. Every question raised by one of the two disciplines corresponds to a question belonging to the other, although neither their methods nor their solutions may coincide.

This independence of methods may be illustrated by a very simple example, whose discussion will moreover be useful to us in what follows (Chapters 5 and 6). It is customary to say that (real) thought "applies the principle of contradiction" which, to take things literally, would mean the intervention of a logical factor in the causal context of psychological facts, and would thus contradict what we have just been asserting. Now, on closer examination of these terms, such a statement is found to be meaningless. The principle of contradiction is confined, in fact, to precluding the simultaneous affirmation and negation of a given predicate: A is incompatible with not-A. But, for the actual thought of a real subject, the difficulty begins when he wonders if he has the right to assert A and B simultaneously, for logic never states directly whether or not B implies not-A. May we, for example, speak of a mountain which is only 100 feet high or is this a contradiction? Is it possible to be both a communist and a patriot? Can we conceive of a square with unequal angles? etc. To answer these questions there are only two possible procedures. The logical procedure consists in formally defining A and B and ascertaining whether B implies not-A. But then the "application" of the "principle" of contradiction relates exclusively to definitions, i.e. to axiomatized concepts and not to the living ideas used by thought in reality. The procedure followed by real thought, on the other hand, consists, not in reasoning on a basis of definitions alone, which has no interest for it (definition being from this point of view only a retrospective and often incomplete act of awareness)—but in acting and operating, in constructing concepts according to the possible combinations of these actions or operations. A concept is in fact only a plan of action or of operation, and only carrying out the operations producing A and B will decide whether they are compatible or not. Far from "applying a principle", actions are organized according to their inner rules of consistency, and it is this organizational structure that constitutes the fact of positive thought corresponding to what is called, on the axiomatic level, the "principle of contradiction."

It is true that in addition to the individual consistency of actions there enter into thought interactions of a collective order and consequently "norms" imposed by this collaboration. But co-operation is only a system of actions, or of operations, carried out in concert, and we may repeat the preceding argument for collective symbolic behaviour, which likewise remains at a level containing real structures, unlike axiomatizations of a formal nature.

For psychology, therefore, there remains unaltered the problem of understanding the mechanism with which intelligence comes to construct coherent structures capable of operational combination; and it is no use invoking "principles" which this intelligence is supposed to apply spontaneously, since logical principles concern the theoretical pattern formulated after thought has been constructed and not this living process of construction itself. Brunschvicg has made the profound observation that intelligence wins battles or indulges, like poetry, in a continuous work of creation, while logico-mathematical deduction is comparable only to treatises on strategy and to manuals of "poetic art", which codify the past victories of action or mind but do not ensure their future conquests.[1]

Meanwhile, and precisely because logical axiomatics schematizes the real work of the mind after it has occurred, every discovery in either of these two fields may give rise to a problem in the other. There is no doubt that logical schemata have by their

---

[1] L. Brunschvicg. *Les Étapes de la philosophie mathématique*, 2nd edition p. 426.

exactness often helped psychological analysis; *Denkpsychologie* is a good example of this. But, conversely, when psychologists like Selz, the "Gestaltists", and many others discover the role of wholes and complex organizations in the work of thought, there is no reason to regard classical logic or even current symbolic logic, which has not gone beyond a discontinuous and atomistic mode of description, as something untouchable and as the last word, or to make of them a model of which thought is the "mirror"; on the contrary, we must construct a logic of wholes if we wish it to serve as an adequate pattern for the states of equilibrium of the mind and to analyse operations without reducing them to isolated and psychologically inadequate elements.

OPERATIONS AND THEIR "GROUPINGS"

The great stumbling-block in the way of any theory of intelligence which starts from the analysis of thought in its higher forms is the fascination that consciousness derives from the ease of verbal thought. P. Janet has shown very ably how language is a partial substitute for action, so that introspection experiences the greatest difficulty in realizing by its own methods that it is itself an item of behaviour; verbal behaviour is an action, doubtless scaled down and remaining internal, a rough draft of action which constantly runs the risk of being nothing more than a plan, but it is nevertheless an action, which simply replaces things by signs and movements by their evocation, and continues to operate in thought by means of these spokesmen. Now, introspection, ignoring this active aspect of verbal thought, sees in it nothing but reflection, speech and conceptual representation, which explains the mistaken belief of introspective psychologists that intelligence is reducible to these privileged terminal states, and the delusion of logicians that the most adequate logistic pattern must be essentially a theory of "propositions".

It is important, therefore, in order to arrive at the real functioning of intelligence, to reverse this natural movement of the
mind and to revert to thinking in terms of action itself; only in
this way will the role of this internal action, the operation,
appear in a clear light. And this very fact forces us to recognize
the continuity which links operation with true action, the source
and medium of intelligence. There is nothing more fitted to
throw light on these facts than a consideration of the sort of
language—still a language, even though it is purely intellectual,
transparent and free from the deceptions of imagery—which we
call mathematics. In any expression, such as $(x^2 + y = z - u)$, each
term refers to a specific action the sign $(=)$ expresses a possible
substitution, the sign $(+)$ a combination, the sign $(-)$ a separation, the square (x^2) the action of reproducing 'x' x times, and
each of the values u, x, y and z the action of reproducing unity a
certain number of times. So each of these symbols refers to an
action which could be realised, but which mathematical language contents itself with describing abstractly in the form of
internalised actions, i.e. operations of thought.[1]

Now if this is obvious in the case of mathematical thought, it
is no less true of logical thought and even of conversational
language from the dual point of view of logical analysis and
psychological analysis. It is in this way that two classes can be
added just like two numbers. In the proposition "Vertebrates and
Invertebrates constitute all the Animals", the word "and" (or the
logical sign $+$) represents an action of combination, which may
be effected materially by classifying a collection of objects but
can also be effected mentally by thought. Similarly, we may

[1] This active character of mathematical reasoning was recognised by Goblot in
his *Traité de logique*; "deduction is construction", he said. But operational construction seemed to him simply to be controlled by the "propositions previously admitted", whereas the control of operations is immanent in them and is
constituted by their capacity for reversible composition, in other words by
their nature as "groups".

make classifications from several points of view at the same time, as in a matrix, and this operation (which symbolic logic calls logical multiplication) denoted by x is so natural to the mind that the psychologist Spearman has gone so far as to make it out to be, under the name of the "education of correlates", one of the distinguishing characteristics of the act of intelligence: "Paris is to France as London is to Great Britain." We may arrange in series the relations A < B; B < C, and this double relation, which permits the conclusion that C is greater than A, is the reproduction in thought of the action which could have been effected materially by placing the three objects in order of increasing size. We may in the same way form series based on several relations at the same time, and come back to another form of logical multiplication or correlation, etc.

But if we now envisage the terms themselves, i.e., the so-called elements of thought class-concepts or relational concepts, we find the same operational character in them as in their combinations. Psychologically a class-concept is only the expression of the identity of the subject's reaction to objects which he combines in one class; logically, this active likening appears as the qualitative equivalence of all the members of the class. Similarly, an asymmetrical relation (more or less heavy or big) expresses different intensities of action, i.e., differences as opposed to equivalences, and in logic takes the form of serial structures.

In short, the essential characteristic of logical thought is that it is operational, i.e., it extends the scope of action by internalising it. On this point we shall unite opinions emanating from the most diverse trends, from empiricist and pragmatist theories, which by attributing the character of a "mental experiment" to thought are forced to accept this basic assumption (Mach, Rignano, Chaslin), to interpretations which are apriorist in inspiration (Delacroix). Furthermore, this hypothesis is in agreement with the schematisations of symbolic logic, as long as these simply devise a technique and are not made into a philosophy denying the existence of the very operations which they are in actual fact constantly using.

However, this does not complete the picture, for an operation is not simply reducible to any and every action, and, even if operational acts are derived from actual acts, the distance between the two is considerable, as we shall see in detail when we come to examine the development of intelligence (Chapters 4 and 5). A rational operation can be compared to a simple action only as long as it is viewed in isolation, but that is precisely the fundamental error of empiricist theories of "mental experiment", that they concentrate on the isolated operation; a single operation is not an operation at all but only a simple intuitive representation. The specific nature of operations, as compared with empirical actions, depends, on the other hand, on the fact that they never exist in a discontinuous state. It is only as an entirely illegitimate abstraction that we speak of "one" operation; a single operation could not be an operation, because the peculiarity of operations is that they form systems. Here we may well protest vigorously against logical atomism, whose pattern has been a grievous hindrance to the psychology of thought. In order to grasp the operational character of rational thought, we must deal with these systems as they are, and, if ordinary logical patterns conceal their existence, then we must construct a logic of wholes.

To begin with the simplest example, classical psychology, like classical logic, speaks in this way of concepts as elements of thought. Quite apart from the fact that its definition relies on other concepts, a "class" could not exist by itself. As an instrument of real thought, disregarding its logical definition, it is only a "structured", not a "structuring", element, or at least it is already structured only in so far as it structures; it has no reality apart from all the entities to which it is opposed or which it includes (or in which it is included). A "class" presupposes a "classification", and the former grows out of the latter, because only operations of classing can engender particular classes. Independently of a general classification, a generic term does not signify a class but an intuitive collection.

Similarly, a transitive, asymmetrical relation, such as A < B, could not exist as a relation (but only as a perceptual or intuitive relationship), were it not for the possibility of constructing a whole succession of serial relations such as A < B < C . . . And when we say that it does not exist as a relation, this denial must be taken in its most concrete sense, because we shall see in Chapter V that the child is in fact quite incapable of thinking in terms of relations before he can serialise. Thus "serialising" is the primary reality, any asymmetrical relation being only an element abstracted from it for the moment.

To take other examples : a "correlate" in Spearmans' sense (dog is to wolf as cat is to tiger) has meaning only as a function of a matrix. A relation of kinship (brother, uncle, etc.) refers to the whole constituted by a family tree, etc. Need we remind the reader that a whole number exists, psychologically as well as logically (in spite of Russell), only by virtue of being an element in the sequence of numbers (engendered by the operation + 1), and likewise that a spatial relation presupposes a whole space, and that a temporal relation implies the conception of time as an exclusive schema? And, in another field, should we insist on the fact that a value is valid only in terms of a complete scale of values, temporary or permanent?

In short, in any possible domain of constituted thought (contrasted with the states of disequilibrium which mark its development), psychological reality consists of complex operational systems and not of isolated operations conceived as elements prior to these systems; thus, only in so far as actions or intuitive representations organise themselves in such systems do they acquire the nature of "operations" (and they acquire it by this very fact). The essential problem of the psychology of thought is then to work out the laws of equilibrium of these systems, just as the central problem of a logic that is to be adequate to the real work of the mind seems to us to be the formulation of the laws governing these wholes as such.

Now, analysis of a mathematical nature has long recognised this interdependence of operations constituting certain well-defined systems; the notion of a "group", which is applied to the series of whole numbers, to spatial or temporal structures, to algebraic operations, etc, has thus become a central idea in the ordering of mathematical thought. In the case of the qualitative systems peculiar to thought that is purely logical, such as simple classifications, matrices, series based on relations, family trees, etc., we shall call the corresponding complex systems "groupings". Psychologically, a "grouping" consists of a certain form of equilibrium of operations, i.e. of actions which are internalised and organised in complex structures, and the problem is to describe this equilibrium both in relation to the various genetic levels which lead up to it and in contrast to forms or equilibrium characteristic of functions other than intelligence (perceptual or motor "structures", etc.) From the logico-mathematical point of view, a "grouping" presents a well-defined structure (related to that of a "group", but differing from it on several essential points), and expressing a succession of dichotomous distinctions; its operational rules thus constitute precisely that logic of wholes which translates into an axiomatic or formal pattern the actual work of the mind when it reaches the operational level of its development, that is to say, its form of final equilibrium.

THE FUNCTIONAL MEANING AND STRUCTURE OF "GROUPINGS"

Let us begin by connecting the foregoing considerations with what we. have learned from "Thought Psychology". According to Selz, the solution of a problem involves in the first place an "anticipatory schema", which links the goal to be attained to a "complex" of ideas in which it creates a gap; then, in the second place, it means the "filling out" of this anticipatory schema by means of concepts and relations which serve to complete the "complex" and are arranged according to the laws of logic. This leads to a series of questions: what are the organisational laws of the total "complex"? What is the nature of the anticipatory schema? Can we abolish the dualism which seems to exist between the formation of the anticipatory schema and the detailed processes which determine the way it is filled out?

By way of example let us take an interesting experiment performed by our colleague, André Rey: a square with sides a few centimetres long is drawn on a sheet of paper which is also square (side 10–15 centimetres), and the subject is instructed to draw with a pencil the smallest square he can as well as the largest square which can be made on such a sheet. Now while adults (and children over the age of 7–8) succeed straight away in producing a square of 1–2 millimetres and one closely following the edges of the paper, children under the age of 6–7 at first draw only squares scarcely smaller and scarcely larger than the standard, and then proceed by successive, and often unsuccessful, trial-and-error, as though they at no time anticipated the final solutions. We can see immediately, in this case, the part played by a "grouping" of asymmetrical relations $(A < B < C . . .)$, which is present in adults and appears to be absent before the age of 7; the perceived square is placed, in thought, in a series of potential squares, becoming bigger and bigger or smaller and smaller in relation to the first. We may then agree:

(i)   that the anticipatory schema is simply the pattern of the grouping itself, that is to say, the consciousness of an ordered series of potential operations.
(ii) that the filling out of the schema is nothing but the putting into practice of these operations.

(iii)   that the organisation of the "complex" of previous ideas obeys the actual laws of grouping.

If this solution is of general validity, the notion of a grouping will thus introduce a unity between the previously existing system of ideas, the anticipatory schema and its controlled filling-out process.

Let us now consider all those concrete problems which the mind in action is continually setting itself: What is it? is it bigger or smaller, heavier or lighter) further or nearer, etc? where? when? what for? to what purpose? how much or many? etc., etc. We note that each of these questions is necessarily dependent on a previous "grouping" or "group"; every individual possesses classifications, seriations, systems of explanation, a personal space and time, a scale of values, etc., as well as mathematical space and time and numerical series. Now these groupings and groups do not come into being when the question is put, but last throughout the individual's life; from infancy onwards, we classify, compare (differences or similarities), locate in space and time, explain, evaluate our ends and our means, count, etc., and problems arise in relation to these total systems just in so far as new facts arise which have not yet been classified, serialised, etc. The question which governs the anticipatory schema thus proceeds from the previously existent grouping, and the anticipatory schema itself is simply the direction imposed on the task by the structure of this grouping. Every problem, whether it concerns the anticipatory hypothesis regarding the solution or its detailed checking, is thus no more than a particular system of operations to be put into effect within the corresponding complex grouping. In order to find our way, we do not have to reconstruct the whole of space, but simply to complete its filling out in a given sector. In order to foresee an event, repair a bicycle, make out a budget or decide on a programme of action, there is no need to build up the whole of causality and time, to review all accepted values, etc.; the solution to be found is attained simply by extending and completing the relationships already grouped, except for correcting the grouping when there are errors of detail, and, above all, subdividing and differentiating it, but not by rebuilding it in its entirety. As for verification, this is possible only in accordance with the rules of the grouping itself, by the fitting of the new relations into the previously existent system.

The remarkable fact in this continuous assimilation of reality to intelligence is, in fact, the equilibrium of the assimilatory frameworks constituted by the grouping. Throughout its formation, thought is in disequilibrium or in a state of unstable equilibrium; every new acquisition modifies previous ideas or risks involving a contradiction. From the operational level, on the other hand, the gradually constructed frameworks, classificatory and serial and spatial, temporal, etc., come to incorporate new elements smoothly; the particular section to be found, to be completed, or to be made up from various sources, does not threaten the coherence of the whole but harmonises with it. Thus, to take the most characteristic example of this equilibrium of concepts, an exact science, despite the "crises" and reforms on which it prides itself to prove its vitality, constitutes a body of ideas whose detailed relationships are preserved and even strengthened with every new addition of fact or principle; for new principles, however revolutionary they may be, justify old ones as first approximations drafted to a certain scale; the continuous and unpredictable work of creation to which science testifies is thus ceaselessly integrated with its own past. We find the same phenomenon again, but on a small scale, in every sane man.

Furthermore, compared with the partial equilibrium of perceptual or motor structures, the equilibrium of groupings is essentially a "mobile equilibrium"; since operations are actions, the equilibrium of operational thought is in no way a state of rest, but a system of balancing interchanges, alterations which are being continually compensated by others. It is the equilibrium of polyphony and not that of a system of inert masses, and it has nothing to do with the false stability which sometimes results in old age from the slowing down of intellectual effort.

It is a question then (and in this lies the whole problem of grouping) of determining the conditions of this equilibrium in order to be able subsequently to examine how it is formed genetically. Now these conditions may be discovered both by observation and by psychological experiment and may be formulated with the degree of precision demanded by an axiomatic pattern. They thus constitute, from the psychological angle, factors of a causal order explaining the mechanism of intelligence, while their logico-mathematical schematisation supplies rules for the logic of wholes.

These conditions are four in number in the case of "groups" of a mathematical order, and five in the case of "groupings" of a qualitative order.

1. Any two elements of a grouping may be combined and thus produce a new element of the same grouping; two distinct classes may be combined into one comprehensive class which embraces them both, two relations A < B and B < C may be joined into one relation A < C which contains them, and so on. Psychologically then, this first condition expresses the possibility of co-ordinating operations.

2. Every change is reversible. Thus, the two classes or the two relations just combined may be separated again and, in mathematical thought, each original operation of a group implies a converse operation (subtraction for addition, division for multiplication, etc.). This reversibility is no doubt the most clearly defined characteristic of intelligence, for although motor functions and perception are capable of combination, they remain irreversible. A motor habit is of a one-way nature, and learning

to effect movements in the other direction means acquiring a new habit. A perception is irreversible since, with each appearance of a new objective element in the perceptual field, there is a "displacement of equilibrium", and since, if we restore the original situation in the outer world, the perception is modified by the intermediate states. Intelligence, on the other hand, can construct hypotheses and then discard them and return to the starting-point, can follow one path and then retrace its steps, without affecting the ideas employed. Now thought in the child, as we shall see in Chapter 5, appears precisely more, irreversible the younger the subject and the nearer to the perceptuo-motor or intuitive patterns of the beginnings of intelligence; reversibility thus characterises not only the final states of equilibrium but also the processes of development themselves.

3. The combination of operations is "associative", (in the logical sense of the term), i.e. thought always remains free to make detours, and a result obtained in two different ways remains the same in both cases. This characteristic seems also to be peculiar to intelligence; perception, like motor functions, is capable only of following one path, since a habit is stereotyped and since, in perception, two distinct paths lead to different results (for example, the same temperature perceived under different conditions of comparison does not seem the same). The appearance of the detour is characteristic of sensori-motor intelligence, and as thought becomes more active and mobile detours play a greater role, but it is only in a system in permanent equilibrium that the final term of the procedure is left constant.

4. An operation combined with its converse is annulled e.g. $+ 1 - 1 = 0$ or $\times 5 \div 5 = \times 1$). On the other hand, in the first forms of thought in the child, the return to the starting-point is not accompanied by a conservation of the latter; for example, having made a hypothesis which he subsequently rejects, the

child does not return to the original data of the problem, because they remain somewhat distorted by the hypothesis, even though it was discarded.

5. In the field of numbers, a unit added to itself yields a new number, by the application of combinativity (1); there is iteration. A qualitative element which is repeated is, however, not transformed; there is a "tautology" in this case: $A + A = A$.

If we express these five conditions of grouping in a logico-mathematical scheme, we arrive at the following simple formulæ:

(I) Combinativity: $x + x^1 = y$; $y + y^1 = z$; etc.
(II) Reversibility: $y - x = x^1$ or $y - x^1 = x$.
(III) Associativity: $(x + x^1) + y^1 = x + (x^1 + y^1) = (z)$.
(IV) General operation of identity:
$x - x = O$; $y - y = O$; etc.
(V) Tautology or special identities:
$x + x = x$; $y + y = y$; etc.

It goes without saying that a calculus of changes becomes possible, but it necessitates, because of the presence of tautologies, a certain number of rules whose details space will not permit us to describe in this book (see Piaget: *Classes, relations et nombres*, Paris, Vrin, 1942).

CLASSIFICATION OF "GROUPINGS" AND OF THE FUNDAMENTAL OPERATIONS OF THOUGHT

The study of the steps in the development of thought in the child leads to the recognition not only of the existence of groupings but also of their mutual connections, i.e. the relations enabling us to classify them and to list them. The psychological existence of a grouping can in fact easily be recognised from the overt operations of which a subject is capable. But that is not all: without the grouping there could be no conservation of complexes or wholes, whereas the appearance of a grouping is attested by the appearance of a principle of conservation. For example, the subject who is capable of reasoning operationally in accordance with the structure of groupings will know in advance that a whole will be conserved independently of the arrangement of its parts, whereas before he would question it. In Chapter 5 we shall study the formation of these principles of conservation in order to show the role of the grouping in the development of reason. But, for clarity of exposition, we had better first describe the final states of equilibrium of thought, so that we may then examine the genetic factors which would explain how they came to be constituted. So, at the risk of producing a rather abstract and schematic enumeration, we shall complete the foregoing remarks by enumerating the principal groupings, it being understood that this sketch represents simply the final structure of intelligence and that the whole problem of understanding their formation still remains unsolved.

I. A first system of groupings is formed by the operations we call logical, i.e., those which start with individual elements which are regarded as constants, and simply classify and serialise them, etc.

1. The simplest logical grouping is that of classification or the formation of hierarchies of more and less inclusive classes. It is based on a primary, fundamental operation: the combining of individuals in classes, and of classes with other classes. The ideal example is found in zoological or botanical classifications, but all qualitative classification follows the same dichotomous pattern.

    Let us suppose that a species A forms part of a genus B, of a family C, etc. The genus B includes other species besides A: we will call them A′ (thus A′ = B − A).

The family C includes other genera beside B: we will call them B′ (thus B′ = C − B) etc. We then have combinativity: A + A′ = B; B + B′ = C; C + C′ = D, etc.; reversibility: B − A′ = A, etc. associativity: (A + A′) − B′ = A + (A′ + B′) = C, etc., and all the other characteristics of groupings. It is this first grouping that gives rise to the classical syllogism.

2. A second elementary grouping brings into play the operation which consists not in combining individuals which are regarded as equivalent (as in 1), but in assembling the asymmetrical relations which express their differences. The linking up of these differences then creates an order of succession and the grouping consequently constitutes a "qualitative seriation":

   Let us call a the relation o < A; b the relation o < B; c the relation o < C. We may then call $a′$ the relation A < B; $b′$ the relation B < C; and we have the grouping: $a + a′ = b$; $b′ + b = c$, etc. The converse operation is the subtraction of a relation, which is equivalent to the addition of its converse. The grouping is parallel to the previous one except for this difference: that the operation of addition implies an order of sequence (and therefore is not commutative). The transitivity peculiar to this serialisation is the basis of the following inference A < B; B < C; therefore A < C.

3. A third fundamental operation is substitution, the basis of the equivalence which joins together the various individuals in a class or the different simple classes included in a composite class:

   Actually, there is not the equality between two elements A1 and A2 of the same class B that there is between mathematical units. There is simply qualitative equivalence, i.e. a possible substitution but only as long as we substitute in the same way as A′1, (i.e. the "other" elements besides

A1) the A'2s (that is the "other" elements besides A2). Hence the groupings: $A1 + A'1 = A2 + A'2$ ($= B$); $B1 + B'1 = B2 + B'2$ ($= C$) etc.

4. Now, interpreted in terms of relations, the preceding operations give rise to the reciprocity which marks symmetrical relations. The latter are, in fact, only the relations uniting the elements of a given class, and therefore they are relations of equivalence (as opposed to asymmetrical relations which denote difference). Symmetrical relations e.g. brother, first cousin, etc.) are consequently grouped after the fashion of the preceding grouping, but each operation is identical with its converse, this being the actual definition of symmetry: $(Y = Z) = (Z = Y)$.

The four preceding groupings are of an additive order, two of them (1 and 3) concerning classes and the other two relations. There exist, in addition, four groupings based on multiplicative operations, i.e. those which deal with more than one system of classes or relations at a time. These groupings correspond, one by one, with the four previous ones

5. Two series of compound classes being given, A1 B1 C1 . . . and A2 B2 C2 . . ., we may start by distributing the individuals according to both systems at once: this is the procedure of matrix tables. Now, "multiplication of classes", which constitutes the characteristic operation of this type of grouping, plays an essential part in the mechanism of intelligence; this is what Spearman describes in psychological terms under the name of "eduction of correlates". The original operation for the two classes B1 and B2 is the product $B1 \times B2 = B1.B2$ ($= A1.A2 + A1.A'2 + A'1.A2 + A'1.A'2$). The converse operation is logical division, $BI.B2 \div B2 = B1$, which corresponds to "abstraction" (B1.B2, disregarding B2, is B1).

6. In the same way we may multiply together two series of relations, i.e. we may find all the relations obtaining among objects serialised according to two sorts of relations at once. The simplest case is none other than qualitative "one-to-one correspondence."

7 and 8. Finally, we may group individuals, not according to the principle of matrices, as in the two previous cases, but by making one term correspond to several, e.g. a father to his sons. In this way, the grouping takes the form of a family tree and is expressed either in classes (7) or in relations (8), the latter thus being asymmetrical in one of its two dimensions (father, etc.) and symmetrical in the other (brother, etc.).

Thus, from the simplest combinations, we obtain eight fundamental logical groupings, some additive $(1 - 4)$, and others multiplicative $(5 - 8)$, some concerning classes and others relations, and some arranged in combinations, seriations or simple correspondences $(1, 2$ and $5, 6)$ and others in reciprocities and correspondences of the "one- many" type $(3, 4$ and $7, 8)$. Hence $2 \times 2 \times 2 = 8$ possibilities altogether.

Further, we should note that the best proof of the natural character of the totalities constituted by these groupings of operations is that it is only necessary to fuse together the groupings formed by simple combination in classes (1) and those formed by seriations (2) in order to obtain what is no longer a qualitative grouping but the "group" constituted by the series of positive and negative whole numbers. In fact, to combine individuals in classes means considering them as equivalent, while serialising them according to an asymmetrical relation expresses their differences. Now, when we consider the qualities of objects we cannot simultaneously group them as both equivalent and different. But if we abstract qualities, by this very fact we render them equivalent to each other and capable of being serialised according to some form of enumeration: we thus transform them into ordered "units", and the additive operation which constitutes a whole number consists in just that. Similarly, by amalgamating multiplicative groupings of classes (5) and relations (6), we obtain the multiplicative group of positive numbers (whole and fractional).

II. The various foregoing systems do not exhaust all the elementary operations of intelligence. Intelligence, indeed, does not confine itself to operating on objects in order to combine them in classes, to serialise or to number them. Its action is also entailed by the construction of the object itself and, as we shall see (Chap. 4), this work has already been completed in the stage of sensori-motor intelligence. Analysing and re-synthesising the object thus constitutes the work of a second type of grouping whose fundamental operations may consequently be called "infralogical", since logical operations combine objects which are regarded as invariant. These infra-logical operations are just as important as logical operations, because they fashion our notions of space and time, whose development occupies almost the whole of childhood. But although quite distinct from logical operations, they are closely parallel to them. The question of the developmental relations between these two operational systems thus constitutes one of the most interesting problems relating to the development of intelligence:

1. Corresponding to the formation of classes is the joining together of parts into progressively more inclusive wholes, whose final term is the whole object (to every possible scale, even that of the spatio-temporal universe itself). It is this first grouping by addition of parts that enables the mind to conceive of atomistic composition prior to any genuinely scientific experiment.

2. Corresponding to seriation by asymmetrical relations are the operations of locating (spatially or temporally) and

qualitative displacement (simple change of order without measurement).

3–4. Spatio-temporal substitutions and symmetrical relations correspond to logical substitutions and symmetries.

5–8. Multiplicative operations simply combine the preceding operations according to several systems or dimensions at once.

Now just as numerical operations may be regarded as expressing a simple fusion of groupings of classes and asymmetrical relations, operations of measurement express the uniting into a single whole of the operations of breaking up into parts and of displacement.

III. We may find the same divisions in the case of operations concerned with values, i.e. those expressing the relations of means and ends which play an essential part in practical intelligence (and whose quantification corresponds to economic value).

IV. Finally, the whole formed by these three systems of operations (I to III) may be interpreted in terms of simple propositions, whence we have a logic of propositions based on implications and contradictions between propositional functions; this is what constitutes logic in the customary sense of the term as well as the hypothetico-deductive theories characteristic of mathematics.

EQUILIBRIUM AND DEVELOPMENT

It has been our purpose in this chapter to find an interpretation of thought which does not clash with logic, regarded as a primary and inexplicable datum, but respects the inherent formal necessity of axiomatic logic, and this while retaining for intelligence its psychological nature as something essentially active and constructive.

Now, the existence of groupings and the possibility of a rigorous axiomatisation of them satisfies the first of these two conditions; the theory of groupings can attain formal precision, even though it arranges systems of logical elements and operations into wholes comparable with the general systems used in mathematics.

From the psychological point of view, on the other hand, since the operations are combinative and reversible actions, but actions nevertheless, continuity between the act of intelligence and all other adaptive processes is still ensured.

But this is merely formulating the problem of intelligence, while the solution still remains to be found. All that the existence and the nature of groupings teach us is that at a certain level thought reaches a state of equilibrium. They tell us, no doubt, what the latter is: an equilibrium, both mobile and permanent, such that the structure of operational wholes is conserved while they assimilate new elements. Further, we know that this mobile equilibrium entails reversibility, which, incidentally, is according to physicists the very definition of a state of equilibrium. (We must conceive the reversibility of the mechanisms of fully developed intelligence in terms of this actual physical model, not in terms of the abstract reversibility of the logico-mathematical pattern). Yet neither pointing out this state of equilibrium nor even stating its necessary conditions constitutes an explanation.

The psychological explanation of intelligence consists in tracing its development and showing how the latter necessarily leads to the equilibrium we have described. From this point of view, the work of psychology is comparable to that of embryology, i.e. a work which, in the first instance, is descriptive and which consists in analysing the phases and periods of morphogenesis up to the final equilibrium constituted by adult morphology, but this study becomes "causal" once the factors which ensure the transition from one stage to the next have been demonstrated. Our task is therefore clear: we must now reconstruct the development of intelligence, or the stages in its formation, until we are able to account for the final operational level whose forms of equilibrium we have just been describing. And since the higher cannot be reduced to the lower—except by distorting the higher or prematurely enriching the lower—the developmental explanation can only consist in showing how, at each new stage, the mechanism provided by the factors already in existence makes for an equilibrium which is still incomplete, and the balancing process itself leads to the next level. In this way, step by step, we may hope to give an account of the gradual formation of operational equilibrium, without having it ready-made from the outset, or having it emerge *ex nihilo* on the way.

Briefly then, the explanation of intelligence amounts to linking the higher operations with the whole process of development, development being regarded as an evolution governed by an inherent need for equilibrium. Now this functional continuity is quite compatible with the differentiation of successive structures. As we have seen, we may represent the hierarchy of response-patterns, right from the early reflexes and global perceptions, as a matter of progressively extending the distances and of progressively complicating the paths of interaction between the organism (subject) and the environment (objects); thus each of these extensions or complications represents a new structure, while their succession is dependent on the need for an equilibrium which must be more and more mobile as it becomes more complex. Operational equilibrium fulfils these conditions on reaching the greatest possible distances (since intelligence tries to embrace the universe) and the greatest possible complexity of paths (since deduction is capable of the greatest "detours"). This equilibrium is therefore to be regarded as the final limit of an evolution whose stages are still to be traced.

Thus the organisation of operational structures goes back far beyond the beginnings of reflective thought and even approaches the origins of action itself. And, since all operations are grouped in well-structured wholes, they must be compared with all "structures", perceptual and motor, are at a lower level. The course to be followed is thus fully sketched; we must analyse the relations between intelligence and perception (Chap. 3) and motor habit (Chap. 4), then we must study the formation of operations in the thought of the child (Chap. 5), and its social-isation (Chap. 6). Only then will the "grouping" structure, which characterises living logic in action, reveal its true nature, whether it be innate or learned (and simply imposed by the environment), or whether it be the expression of ever more numerous and complex interactions between subject and objects, interactions at first incomplete, unstable and irreversible, but gradually acquiring, by the very needs arising from the equilibrium which is forced on them, the form of reversible combinativity characteristic of the grouping.

Part II

Intelligence and Sensori-motor
Functions

3

INTELLIGENCE AND PERCEPTION

Perception is the knowledge we have of objects or of their movements by direct and immediate contact, while intelligence is a form of knowledge obtaining when detours are involved and when spatio-temporal distances between subject and objects increase. It is possible then that intellectual structures, and notably the operational groupings which characterise the final equilibrium reached in the development of intelligence, pre-exist, wholly or in part, from the outset in the form of organisations common to perception and to thought. This particular idea is the central doctrine of the "Configuration theory", which, although it knows nothing of the notion of a reversible grouping, has described laws of complex structuring which, it claims, govern perception, response and elementary functions as well as reasoning itself and in particular the syllogism (Wertheimer). It is therefore essential that we should start with perceptual structures, to enquire whether we may not derive from them an explanation of the whole of thought, including groupings themselves.

HISTORICAL

The hypothesis of a close relationship between perception and intelligence has been maintained at all times by some, and likewise rejected at all times by others. We shall mention here writers of experimental studies only, as opposed to the innumerable philosophers who have confined themselves to "reflecting" on the subject. And we shall set forth the point of view of experimenters who have sought to explain perception by the intervention of intelligence as well as that of those who seek to derive the latter from the former.

It was undoubtedly Helmholtz who first framed the problem of the relations between perceptual structures and operational structures in its modern form. We know that visual perception can show "constancy" effects which have stimulated and still stimulate a series of studies. A given size is perceived more or less correctly at a distance, in spite of the considerable contraction of the retinal image and the diminution due to perspective; a shape is recognised even at an angle; colour is recognised in the shadow as well as in bright illumination, etc. Now Helmholtz tried to explain these perceptual constancies by the intervention of an "unconscious inference" which has the effect of correcting the immediate sensation by recourse to acquired knowledge. When we recall Helmholtz's preoccupations with the formation of the notion of space, we can well imagine that this hypothesis was bound to have a certain significance in his thought, and Cassirer assumed (when he in his turn took up the idea) that the great physiologist, physicist and geometer tried to account for perceptual constancy by a sort of geometrical "group", inherent in the intelligence which works unconsciously in perception. Now, at this stage, it is very interesting for comparative purposes to examine some intellectual and perceptual mechanisms. Perceptual "constancy" is, in fact, comparable, at the sensori-motor level, with the various ideas of "conservation" which characterise the first conquests of intelligence (conservation of wholes, of substance, of weight, of volume, etc. occurring with intuitive distortions). Now, since these ideas of conservation are due to the intervention of a "grouping" or a "group" of operations, if visual constancy were itself due to unconscious inference in the form of a "group", there would then be a direct structural continuity between perception and intelligence.

However, Hering had already replied to Helmholtz, indicating that the fact of intellectual knowledge does not modify a perception; we demonstrably experience the same optical or weight illusion etc. when the objective values of the perceived material are known. He therefore concluded that reasoning is not involved at all in perception and that "constancy" is due to purely physiological regulations.

But Helmholtz and Hering both believed in the existence of sensations that were prior to perception, and so they thought of perceptual constancy as a correction of sensations, and attributed it, in the case of Helmholtz, to intelligence and, in the case of Hering, to neural mechanisms. The problem was revived after von Ehrenfels in 1891 discovered the perceptual qualities of wholes (*Gestaltqualitäten*), such as that of a melody, which can be recognised despite a transposition that changes every note (so that no elemental sensation remains the same). Two schools arose as a result of this discovery, one of them supporting Helmholtz in his appeal to intelligence. The Graz school (Meinong, Benussi, etc.) continue, in fact, to believe in sensations and accordingly interpret a "whole quality" as the product of a synthesis; this synthesis, being transposable, is conceived as something due to intelligence itself. Meinong has gone so far as to build up on the basis of this interpretation a whole theory of thought based on the idea of a whole (the "collective objects" linking the perceptual and the conceptual). On the other hand, the "Berlin school", which marks the starting point of Gestalt Psychology has reversed the position; for this school, sensations no longer exist as elements prior to perception or independent of it (they are "structured" instead of "structuring contents"), and the total configuration, a concept applied generally to all perception, is no longer regarded as the result of a synthesis but as a primary fact produced unconsciously and as much physiological in nature as psychological. These "configurations" (*Gestalten*) are met with at every stage of the mental hierarchy and, according to the Berlin school, we may therefore expect an explanation of intelligence which starts from perceptual structures, instead of assuming that, in some incomprehensible manner, reasoning intervenes in perception itself.

Among later researches, a school known as the *Gestaltkreis* (of Weizsäcker, Auersperg, etc.) has tried to extend the idea of a complex structure by regarding it as embracing perception and bodily movement from the outset, believing these to be of necessity closely associated. Perception would then involve the intervention of motor anticipations and reconstitutions which, without implying intelligence, nevertheless presage it. So we may consider this trend as a revival of the Helmholtzian tradition, while other contemporary studies adhere to Hering's suggestion of an interpretation of perception in purely physiological terms (Piéron, etc.).

THE GESTALT THEORY AND ITS INTERPRETATION OF INTELLIGENCE

Special mention must be given to the Gestalt point of view, not only because it has raised a large number of problems anew, but especially because it has provided a complete theory of intelligence which will remain, even for its opponents, a model of coherent psychological interpretation.

The central idea of the Gestalt theory is that mental systems are never constituted by the synthesis or association of elements that exist in isolation before they come together, but always, from the outset, consist of organised wholes in a "configuration" or complex structure. Thus, a perception is not the synthesis of previous sensations; it is governed at each level by a "field" whose elements are interdependent by the very fact that they are perceived as a whole. For example, a single black dot seen on a large sheet of paper could not be perceived as an isolated element, although it is quite alone, since it stands out as a "figure" on a "ground" formed by the paper, and since this figure-ground relation implies an organisation of the entire visual field. The truth of this is emphasised by the fact that, strictly, one should be able to perceive the sheet as the object (the "figure") and the black dot as a whole, i.e. as the only visible part of the "ground". Why then do we prefer the first mode of perception? And if, instead of a single dot, we see three or four fairly close together, why is it that we cannot help forming them into potential shapes as triangles or quadrilaterals? It is because elements perceived in the same field are immediately bound together in complex structures in accordance with precise laws, i.e., the "laws of organisation".

These laws of organisation governing all the relations within a field are, according to the "Gestalt" hypothesis, simply the laws of equilibrium governing the neural excitation released both by psychological contact with external objects and by the objects themselves, combined in a closed circuit which embraces the organism and its immediate environment simultaneously. From this point of view, a perceptual (or motor, etc.) field is comparable to a field of forces (electro-magnetic, etc.,) and is governed by analogous principles of minima, or of least action, etc. Faced with a multiplicity of elements, we impress upon them a complex pattern, which is not just any pattern but the simplest possible pattern which expresses the structure of the field; so this involves rules of simplicity, regularity, proximity, symmetry, etc. which will determine what configuration will be perceived. Hence we have an essential law (called *Prägnanz*): out of all possible configurations, the configuration which predominates is always the "best", i.e. the best equilibrated. Moreover, a "good Gestalt" is always capable of being "transposed", like a melody when all the notes are changed. But this transposition, which demonstrates the independence of the whole with respect to the parts, is also explained by laws of equilibrium; the same relations hold between the new elements, which give rise to the same total configuration as the old elements, not because of an act of comparison but by means of a reestablishment of equilibrium, in the same way as canal water keeps the same horizontal form, although at different levels, as each sluice-gate is opened. The description of these "good *Gestalten*" and the study of these "transpositions" have given rise to a host of experimental studies of undeniable interest, the details of which it would not help to describe here.

On the other hand, it must be carefully noted as essential to the theory that the "laws of organisation" are considered to be independent of development and consequently common to all stages. This statement follows automatically if we confine it to functional organisation or "synchronous" equilibrium of behaviour, because the necessity for the latter operates at all stages, whence arises the functional continuity on which we have insisted. But it is customary to make a distinction between this constant functioning and successive structures considered from a "diachronic" point of view, which vary from one stage to another. The distinguishing mark of the Gestalt School is that it combines function and structure into one whole under the name of "organisation", and regards the laws of the latter as invariable. In this way Gestalt psychologists have striven to show, with an impressive accumulation of material, that perceptual structures are the same in the young child and the adult and, in fact, that they are the same in vertebrates of all types. The only point of difference between child and adult might be the relative importance of certain common factors of organisation—e.g.

proximity—but the mass of factors remains the same and the resulting structures obey the same laws.

In particular, the famous problem of perceptual constancy has yielded a systematic solution, concerning which the following two points should be noticed. In the first place, constancy such as that of size could not consist in the correction of an initial distorting sensation associated with a diminished retinal image, because no initial isolated sensation exists, and because the retinal image is only a link (and not an especially privileged one) in the chain, whose closed circuit links objects with the brain through the medium of the neural processes involved. Thus, when an object is seen at a distance, its real size is immediately and directly perceived, simply by virtue of the laws of organisation which make this the best of all structures. In the second place, therefore, perceptual constancy is held not to be acquired but to be completely formed at all levels, in the animal and the infant just as in the adult. The apparent experimental exceptions would be due to the fact that the "perceptual field" is not always sufficiently structured, the best constancy occurring when the object forms part of a complex configuration, such as a succession of objects forming a series.

To turn back to intelligence, it has received, from this point of view, a remarkably simple interpretation and one which, if it were true, would be capable of establishing an almost complete connection between higher structures (and especially the "operational groupings" we have described) and the most elementary "configurations" of a sensori-motor or even perceptual order. Three applications of the Gestalt theory to the study of intelligence are especially noteworthy: that of Köhler to sensori-motor intelligence, that of Wertheimer to the structure of the syllogism, and that of Duncker to the act of intelligence in general.

For Köhler, intelligence appears when perception is not carried over directly into responses likely to ensure the attainment of the objective. A chimpanzee in a cage tries to reach a fruit placed beyond the reach of his arm. Thus an intermediate agent is required, whose use will constitute the definition of the degree of complication characteristic of intelligent behaviour. What does this consist of? If a stick is placed within reach of an ape but in any position, it is seen as an indifferent object; placed parallel with his arm, it will promptly be perceived as a possible extension of the hand. Thus the stick, until then neutral, will receive a meaning from the fact of its incorporation in the complex structure. The field will then be "restructured" and, according to Köhler, it is these sudden restructurings that are characteristic of the act of intelligence. The shift from a less good structure to a better structure is the essence of insight and is consequently a simple but mediate or indirect continuation of perception itself.

This is the explanatory principle that occurs again in Wertheimer's Gestalt interpretation of the syllogism. The major term is a *"Gestalt"* comparable to a perceptual structure; "all men" thus constitutes a whole which is represented as located within the complex of "mortals". The minor term follows the same course; "Socrates" is an individual located within the circle of "men". So the operation which draws the conclusion from these premises, "therefore Socrates is mortal", simply amounts to restructuring the whole by abolishing the intermediate circle (men) after first placing it, with its content, within the large circle (mortals). Reasoning is thus a "re-anchoring". "Socrates" is, so to speak, uprooted from the class of "men" in order to be anchored in that of "mortals". The syllogism is thus without more ado related to the general organisation of structures; in this it is analogous to the restructurings that characterise Köhler's practical intelligence, but it now takes place in thought, not in action.

Finally, Duncker studied the relation of these sudden insights (*Einsicht* or intelligent restructuring) to past experience and so dealt the *coup de grâce* to associationist empiricism, which the concept of a *Gestalt* opposes from its very origins. To this end, he analyses various problems of intelligence and finds in all cases that past experience plays only a secondary role in reasoning; experience never introduces meaning into thought except as a function of present organisation. It is the latter (i.e. the structure of the present field) that determines what appeals to past experience can be made, whether it makes them useless or whether it commands the summoning up and utilisation of memories. Reasoning is thus "a contest which contrives its own weapons", and all this is explained by the laws of organisation, which are independent of the individual's history and, in short, ensure the fundamental unity of the structures of every level, from elementary perceptual "configurations" to those of the most exalted thought.

CRITIQUE OF GESTALT PSYCHOLOGY

We are bound to admit how well founded are the descriptions given by Gestalt psychology. The essential "wholeness" of mental structures (perceptual as well as intelligent), the existence of the "good *Gestalt*" and its laws, the reduction of variations of structure to forms of equilibrium, etc., are justified by so many experimental studies that these concepts have acquired the right to be quoted throughout contemporary psychology. In particular, the method of analysis that consists in always interpreting facts in terms of a total field is alone justifiable, since reduction to atomistic elements always impairs the unity of reality.

But it is as well to recognise that, if the "laws of organisation" are not derived, beyond psychology and biology, from absolutely general "physical *Gestalten*" (Köhler),[1] then the language of

---

[1] According to Köhler, "physical Gestalten" have the same role in relation to mental structures that eternal Ideas have in relation to concepts according to Russell, or that an *a priori* framework has in relation to living logic.

wholes is merely a mode of description, and the existence of total structures requires an explanation which is not at all included in the fact of wholeness. We have admitted this in connection with our own groupings and we must also admit it in connection with "configurations" or elementary structures.

The general and even "physical" existence of "laws of organisation" implies at the very least—and Gestalt theorists are the first to vouch for it—their constancy in the course of mental development. The essential question for the orthodox Gestalt doctrine (we shall adhere to this orthodoxy for the moment, but we must point out that certain of the more cautious partisans of the Gestalt school, such as Gelb and Goldstein, have rejected the hypothesis of "physical *Gestalten*") is thus that of the permanence of certain essential forms of organisation throughout mental development, e.g. that of perceptual constancy.

However, as far as the main point is concerned, we think it is possible to maintain that, in the present state of knowledge, the facts are opposed to such an assertion. Without going into detail, and confining ourselves to the field of child psychology and size constancy, we must now consider the following few points:

1. H. Frank[1] believed that he could demonstrate size constancy in infants of 11 months. Now the technique of these experiments has evoked discussion (Beyrl) and, even if the facts are on the whole correct, 11 months already represents a considerable development of sensori-motor intelligence. E. Brunswik and Cruikshank have noted a progressive development of this constancy during the first six months.

2. Certain experiments, consisting of paired comparisons of heights at a distance, which the author has carried out in collaboration with Lambercier on children aged 5–7 years, have enabled us to illustrate a factor which experimenters had not

[1] *Psychol Forschung* VII, 1926, pp. 137–154.

taken into account: at every age there exists a "systematic error of the standard" whereby the element chosen as the standard, for the very reason that it functions as a standard, is over-estimated in relation to the variables that are measured by it, both when it is placed at a distance and when it is near. This systematic error on the part of the subject, combined with his estimations at a distance, could give rise to an apparent and illusory constancy. Calculation of the "error of the standard" shows our 5-7 year old subjects to have moderately under-estimated under conditions of depth perception, whilst adults tend, on the average, towards a "superconstancy".[1]

3. Burzlaff,[2] who has also obtained variations with age in paired comparisons, has considered it possible to maintain the Gestalt hypothesis of the permanence of size constancy in the case where the compared elements are enclosed in a total "con-figuration" and especially when they are serialised. With some painstaking experiments, Lambercier, at our request, has taken up this problem of serial comparisons in depth perception,[3] and has been able to show that a constancy that was relatively independent of age existed only in a single case (the very one that Burzlaff expected): the case where the standard equals the median of the compared elements. On the other hand, as soon as a standard is chosen that is appreciably larger or smaller than the median, systematic changes with distance are observed. Hence it is clear that the constancy of the median depends on other causes than constancy with distance; it is its privileged position as the median that ensures its invariability (it is reduced by all higher terms and correspondingly increased by all lower terms: hence its stability). Again, measurements of the other terms show that specific constancy with distance does not exist in the

[1] *Arch. de Psychol.* XXIX (1943). pp. 255–308.

[2] *Zeitschr. Für Psychol.*, vol. 119 (1935), pp. 177–235.

[3] *Arch. de Psychol.* XXXI (1946).

child, while a remarkable growth with age of the regulations conducive to this constancy is observed.

4. We know that when Beyrl[1] analysed size constancy in school-children, he for his part found some increase in the incidence of constancy up to nearly ten years of age; beyond this stage the child comes to react in the adult manner (a parallel development was found by Brunswik with respect to shape and colour constancy).

The existence of a development with age of the mechanisms underlying perceptual constancy (and later we shall see many other developmental changes in perception) undoubtedly leads to a revision of the Gestalt School's explanation. To begin with, if there is an actual development of perceptual structures, we can no longer dismiss either the problem of their formation or the possible role of past experience in the process of their coming into being. Concerning this last point, Brunswik has demonstrated the frequency of empirical *Gestalten* side by side with "geometrical *Gestalten*". In this way, a figure that is intermediate between the image of an open hand and a geometrical pattern with five exactly symmetrical extensions, when seen tachistoscopically, yielded in adults 50 per cent in favour of the hand (learned shape) and 50 per cent in favour of the geometrical "good *Gestalt*".

Concerning the genesis of *Gestalten*, which raises an essential problem as soon as we reject the hypothesis of permanent "physical *Gestalten*", we may first of all point out the illicit nature of the dilemma: either wholes or the atomism of isolated sensations. In point of fact there are three possible terms. A perception may be a synthesis of elements, or else it may constitute a single whole, or it may be a system of relations (each relation being itself a whole, but the complete whole becoming unanalysable and not relying at all on atomism). This being the case, there is

---

[1] *Zeitschr. für Psychol.*, vol. 100 (1926), pp. 344–371.

no reason why complex structures should not be regarded as the product of a progressive construction which arises, not from "syntheses", but from adaptive differentiations and combined assimilations, nor is there any reason why this construction should not be related to an intelligence capable of genuine activity as opposed to an interplay of pre-established structures.

With regard to perception, the crucial point is that of "transposition". Should we follow Gestalt theory and interpret transpositions (of a melody from one key to another or of a visual form by enlargement) as the simple reappearance of the same form of equilibrium between new elements whose relations have been retained (cf. the horizontal levels of systems of sluice-gates), or should we regard them as the product of an assimilatory activity which integrates comparable elements into the same schema? The fact of improvement with age in ability to transpose (see the end of this chapter) seems to us to demand this second solution. Moreover, transposition as ordinarily understood, which is external to the figures, should undoubtedly be connected with the internal transpositions between elements of the same figure, which explain the role of the factors of regularity, equality, symmetry, etc., inherent in "good *Gestalten*".

These two possible interpretations of transposition mean quite different things with respect to the relations between perception and intelligence and especially the nature of the latter.

In attempting to reduce the mechanisms of intelligence to those characterising perceptual structures, which are in turn reducible to "physical *Gestalten*", the Gestalt theory reverts essentially to classical empiricism, although by far more refined methods. The only difference (and considerable though it is, it has little weight in the face of such a reduction) is that the new doctrine replaces "associations" by structured "wholes". But in both cases operational activity in sensory processes fades into pure receptivity, and abdicates in favour of the passivity of automatic mechanisms.

We cannot insist too strongly on the fact that, although operational structures are bound to perceptual structures by a continuous series of intermediate structures (and we grant this without any difficulty), there is, nevertheless, a fundamental contradiction in meaning between the rigidity of a perceived "configuration" and the reversible mobility of operations. Thus Wertheimer's attempted comparison between the syllogism and the static "configurations" of perception runs the risk of remaining inadequate. What is essential in the mechanism of a grouping (by which syllogisms are formed) is not the structure assumed by premises, nor that which characterises conclusions, but rather the process of combination which makes it possible to pass from the one to the other. No doubt this process is an extension of perceptual restructurings and recentrings (such as those enabling us to see an "ambiguous" design alternately as convex and concave). But it is even more than this, since it is constituted by the whole system of mobile and reversible operations of conjunction and disjunction ($A + A' = B$; $A = B - B'$; $A' = B - A$; $B - A - A' = O$, etc.) So it is no longer static forms that are important in intelligence, nor the simple unidirectional transition from one state to another (or even oscillation between the two); the general mobility and reversibility of operations are what give rise to structures. It follows that the structures involved themselves differ in the two cases. A perceptual structure is characterised, as the Gestalt theory itself has insisted, by its irreducibility to additive combination—it is thus irreversible and non-associative. So there is considerably more in a system of reasoning than a "recentring" (*Umzentrierung*); there is a general decentralisation, which means a dissolution or melting down of static perceptual forms in favour of operational mobility, and consequently there is the possibility of constructing an infinite number of new structures which may be perceptible or may exceed the limits of all true perception.

As for the sensori-motor intelligence described by Köhler, it is clear that here perceptual structures play a much bigger part. But by the very fact that Gestalt theory is bound to consider them as arising directly from situations as such, without historical development, Köhler found himself constrained to exclude from the realm of intelligence, on the one hand, the trial-and-error which precedes the discovery of solutions and, on the other hand, the corrections and checks which follow it. Study of the child's first two years of life has led us, in this context, to a different viewpoint. There are indeed complex structures or "configurations" in the infant's sensori-motor intelligence, but far from being static and non-historical, they constitute "schemata" which grow out of one another by means of successive differentiations and integrations, and which must therefore be ceaselessly accommodated to situations by trial-and-error and corrections at the same time as they are assimilating the situations to themselves. The response with the stick is thus prepared by a series of anticipatory schemata, such as that of pulling the objective to oneself by means of its extensions (string or struts) or that of striking one object against another.

The following reservations must therefore be made with respect to Duncker's thesis. An act of intelligence is doubtless determined by past experience only in so far as it resorts to it. But this relationship involves assimilatory schemata which in turn are the product of previous schemata, from which they are derived through differentiation and co-ordination. Schemata thus have a history; there is interaction between past experience and the present act of intelligence, not uni-directional action of past on present as empiricism demands nor uni-directional appeal to the past by the present as Duncker would have it. It is even possible to formulate these relations between present and past by saying that equilibrium is reached when all previous schemata are embedded in present ones and intelligence can equally well reconstruct past schemata by means of present ones and vice versa.

On the whole then, we see that the Gestalt theory, although correct in its description of forms of equilibrium or well-structured wholes, nevertheless neglects the reality, in perception as in intelligence, of genetic development and the process of construction that characterises it.

DIFFERENCES BETWEEN PERCEPTION AND INTELLIGENCE

The Gestalt theory has revived the problem of the relations between intelligence and perception by demonstrating the continuity which links the structures characteristic of these two fields. The fact remains that, in order to solve the problem while respecting the complexity of genetic facts, we must list the differences between them before considering analogies leading to possible explanations.

A perceptual structure is a system of interdependent relations. Whether geometrical forms are involved, or weights, or colours, or sounds, the wholes can always be interpreted in terms of relations without destroying the unity of the whole as such. For the purpose of distinguishing the differences as well as the similarities between perceptual and operational structures, it is sufficient to express these relations in terms of "groupings", just as physicists, when they formulate the phenomena of thermodynamics in reversible terms, prove that they cannot be interpreted in such terms because they are irreversible. The non-correspondence of symbolic systems thus emphasises all the more the differences involved. In this respect, it is sufficient to reconsider the various well-known geometrical illusions and to vary the factors present, or the facts relating to Weber's law, etc., and to formulate all the relations in "grouping" terms and their changes as a function of external modifications.

Now the results thus obtained have made themselves clear. None of the five conditions of "grouping" is realised at the level of perceptual structures, and where they seem to come nearest to being realised, as in the case of "constancies", which herald operational conservation, the operation is replaced by simple regulations which are not entirely reversible (and consequently midway between spontaneous irreversibility and operational control).

As a first example, let us take a simplified form of Delbœuf's illusion.[1] A circle A1, of radius 12 mm., drawn within a circle B of 15 mm., appears larger than an isolated circle, A2, equal to A1. We vary the external circle B by altering its radius gradually from 15 to 13 mm., and from 15 to 40 or 80 mm.: the illusion is reduced from 15 to 13 mm.; it is also reduced from 15 to 36 mm., reaching zero at about 36 mm., (i.e. when the diameter of A1 equals the width of the zone between B and A1) and becoming negative beyond this point (under-estimation of the inner circle A1). Now:

1. If we are to translate the relations occurring in these perceptual changes into operational language, it is obvious straight away that their combination could not be additive, because the conservation of the elements of the system is lacking. Furthermore, this is the essential discovery of the Gestalt theory and, according to the theory, characterises the idea of perceptual "wholeness". If we call A′ the intervening area marking the difference between the circles A1 and B1, we could not write A1 + A′ =B, since A1 is distorted by its insertion in B, since B is distorted by the fact of surrounding A1, and since zone A′ is more or less expanded or contracted according to the relations between A1 and B. We may prove this non-conservation of the whole in the following manner. If, starting from a certain value of A1, A′ and B, we enlarge (objectively) A1, thus reducing A′

<hr />

[1] See Piaget, Lambercier, etc., *Arch. de Psychol.*, XXIX (1942), pp. 1–107.

but leaving B constant, it is possible that the whole of B will appear smaller than before. It will thus have lost something during the change; or, on the other hand, it will appear larger and something extra will have been added. The problem then is to find a means of formulating these "uncompensated changes".

2.   With this aim in view, let us interpret the changes in terms of the combination of relations and we shall demonstrate the irreversible nature of this combination, this irreversibility expressing in another form the absence of additive combination. We will call s the increase in dimensional similarity between A1 and B, and d the increase in dimensional difference between the same terms. These two relations are bound to be and to remain the converse of each other: $+s = -d$ and $+d = -s$ (the sign $-$ indicating the decrease in similarity or difference). Now, if we start with no illusion (A1 = 12 mm. and B = 36 mm.), we find that as objective similarity is increased (by compressing the circles) the subject perceives it to be still more reinforced. Consequently, perception has increased similarity to excess when it was objectively increased, and inadequately maintained the difference when it was objectively reduced. Similarly, when the objective difference is increased (by widening the circles), this increase is also exaggerated. There is thus a lack of compensation in the course of the transformations. So we may agree to set out these transformations in the following form, which is intended to denote their non-combinative character from a logical standpoint:

$$s > -d \text{ or } d > -s$$

In fact, if in each figure considered separately the relations of similarity are automatically always the converse of the relations of difference, the sum of the similarities and differences will not remain constant with transition from one figure to another, since the wholes are not conserved (see under 1). This is the sense in which we may legitimately regard increases in similarity as outweighing decreases in difference or vice versa.

It is possible, in this case, to express the same idea more concisely simply by saying that the change in the relations is irreversible because it is associated with all "uncompensated change" P, such that:

$$s = -d + P_{sd} \text{ or } d = -s + P_{sd}$$

3. Moreover, no combination of perceptual relations is independent of the route travelled to reach it (associativity), but each perceived relation depends on those which immediately preceded it. Thus, the perception of the same circle A will yield palpably different results according to whether it is compared with reference circles arranged in ascending or in descending order. In this instance, the most objective measure is a random order, that is to say, one which employs sometimes larger and sometimes smaller elements than A, so that they compensate each other for the distortions due to previous comparisons.

4 and 5. It is therefore obvious that a given element does not remain the same when compared with others different from it and when it is compared with others of the same dimensions as itself. Its value will continually vary as a function of the relations given, present as well as past.

So it is impossible to reduce a perceptual system to a "grouping", except by turning inequalities into equalities by introducing "uncompensated changes" P, which measure the extent of distortion (illusions) and bear witness to the non-additivity and non-transitivity of perceptual relations, to their irreversibility, to their non-associativity and to their non-identity.

This analysis (which incidentally gives us some idea of what thinking would be like if its operations were not "grouped"!) shows that the form of equilibrium inherent in perceptual structures is quite different from that of operational structures. In the latter, equilibrium is both mobile and permanent, and changes within the system do not modify it because they are always exactly compensated, owing to actual or potential converse operations (reversibility). In the case of perception, on the other hand, each modification of the value of one of the relations involved means a change of the whole, to the extent of introducing a new equilibrium distinct from the one characterising the previous state. There is then "displacement of equilibrium" (as they say in physics in connection with the study of irreversible systems as in thermodynamics), and no longer permanent equilibrium. This is the case, for example, for each new value of the outer circle B, in the illusion just described. The illusion increases or diminishes but does not conserve its original value.

Moreover, these "displacements of equilibrium" obey laws of maxima; a given relation generates an illusion and so produces an uncompensated change P, as judged by the value of other relations, only up to a certain value. Beyond this value the illusion diminishes, because the distortion is then partially compensated by the effect of the new relations of the whole. So displacements of equilibrium give place to regulations or partial compensations, which may be defined as the change of sign of the quantity P (e.g. when the two concentric circles are too close or too far apart, Delboeuf's illusion is reduced). Now these regulations, the effect of which is thus to limit or "restrict" (as they say in physics) the displacements of equilibrium, are comparable in certain respects to the operations of intelligence. If the system were of an operational order, every increase in one of the values would correspond to a decrease in another and vice versa (there would then be reversibility, i.e. $P = 0$); if, on the other hand, there were unlimited distortion with every external modification, the system would no longer exist as such; the existence of regulations thus manifests the existence of an intermediate structure between complete irreversibility and operational reversibility.

But how are we to explain this relative opposition (paralleled by a relative affinity) between perceptual and intelligent mechanisms? The relations which compose a total structure such as that of a visual perception express the laws of a subjective or perceptual space, which may be analysed and compared with geometrical space or operational space. Illusions (or uncompensated changes in the system of relations) may now be conceived as distortions of this space in the direction of expansion or contraction[1].

According to this point of view, one essential fact governs all relations between perception and intelligence. When intelligence compares two terms with each other, as in measuring one by means of the other, neither the standard nor the compared entity (in other words, neither the measure nor what is measured) is distorted by the comparison itself. On the other hand, in the case of perceptual comparison, and especially when one element acts as a fixed standard for the evaluation of variable elements, a systematic distortion is produced which we, in company with Lambercier, have called the "error of the standard". The element which is fixated most (i.e. generally the standard itself when the variable is at a distance from it but also sometimes the variable when the standard is close to it and already known) is systematically over-estimated, and this applies to comparisons made in the frontal parallel plane as well as in depth[2].

[1] Thus, in Delbœuf's illusion, the area of the inner circle A1 appears expanded at the expense of that of the zone A' between this circle and the outer circle B, when this zone A' is narrower than the diameter of A1; when A' > A1 the effect is reversed.

[2] The proof that it is really a question of an error bound up with the functional status of the measure is that this error can be reduced, or even abolished, by pretending to change the standard for each comparison (while actually retaining the same one throughout). The perceptual error may even be reversed by causing the verbal judgment to be made on the standard instead of on the measured stimulus (if the subject says A < B we require the judgment B > A), which reverses the functional positions.

Such facts as these merely constitute particular cases of a very general process. If the standard is over-estimated (or, in certain circumstances, the variable) it is simply because the element which is fixated longest (or most often, or most intensely, etc.), is by this very fact magnified, as though the object or the region on which vision is concentrated, occasioned an expansion of perceptual space. In this respect, we have only to fixate two equal elements successively to see that on each occasion the dimensions of the one fixated are enhanced, although, taken as a whole, these successive distortions compensate each other. Perceptual space then is not homogeneous but is centralised from moment to moment, and the area of centralisation corresponds to a spatial expansion, while the periphery of this central zone is progressively contracted as one proceeds outwards from the centre. This role of centralisation and of the error of the standard is found also in the tactile sense.

But although "centralisation" thus causes distortions, several distinct centrings correct one another's effects. "Decentralisation", or co-ordination of different centrings, is consequently a correcting factor. So we see straight away the rudiments of a possible explanation for irreversible distortions and for the regulation we have just been discussing. Illusions of visual perception may be explained by the mechanism of centralisation when the elements of the figure are (relatively) too close to each other for decentralisation to occur (illusions of Delbœuf, Oppelkundt, etc.). Conversely, regulation occurs to the extent that there is decentralisation, either automatic or by active comparison.

We see now the relationship between these processes and those characterising intelligence. It is not only in the field of perception that (relative) error is associated with centralisation and (relative) objectivity with decentralisation. The whole of the development of thought in the child, the initial intuitive forms of which are closely related to perceptual structures, is characterised by a transition from a general egocentricity (which we shall reconsider in Chapter 5) to intellectual decentralisation, and thus by a process comparable to the one whose effects we are here ascertaining. But, for the moment, the problem is to understand the differences between perception and complete intelligence and, in this respect, the foregoing facts enable us to grasp more fully the chief of these contrasts: the contrast between what might be called "perceptual relativity" and intellectual relativity.

Indeed, if centrings are interpreted as distortions which, as we have seen, may be formulated by reference to (and by contrast with) a grouping, the next problem is to measure them as far as is possible and to interpret this quantification. This may conveniently be done in the case where two homogeneous elements are compared with each other, as in the case of two straight lines which are extensions of each other. We may then state a law of "relative centralisation" which is independent of the absolute value of the effects of centring, and expresses relative distortions in the form of a single probable value, i.e. by the relation of actual centrings to the number of possible centrings.

We know that a line A, compared with another line A', is underestimated if the second is larger than the first $(A < A')$ and overestimated in the opposite case $(A > A')$. The method of calculation is, in each of these two cases, to consider the successive centralisations on A and on A' as alternately enlarging these lines in proportion to their lengths. The difference of these distortions, expressed in relative sizes of A to A', thus gives the gross overestimation or under-estimation of A. These are then divided by the total length of the contiguous lines $A + A'$, since the decentralisation is proportional to the size of the total figure. We then obtain:

$$\frac{(A-A')\,A'/A}{A+A'} \text{ and where } A > A' \text{ and } \frac{(A'-A)\,A/A'}{A+A'} \text{ where } A < A'.$$

Furthermore, if the measurement is made on A, these relations must be multiplied by $A^2/(A + A')^2$, i.e. by the square of the ratio of the part measured to the whole.

The theoretical curve obtained in this way corresponds closely to empirical measurements of distortions and, moreover, coincides fairly accurately with measurements of Delbœuf's illusion[1] (if, in the formula, A is inserted between the two A's and this value A' is doubled).

Expressed in qualitative language, this law of relative central-isation simply means that every objective difference is subject-ively accentuated by perception, even in the case where the compared elements are equally centred in vision. In other words, all contrast is exaggerated by perception, which immediately points to the presence of a relativity peculiar to the latter and distinct from the relativity of intelligence. This brings us to Weber's law, the discussion of which in this context is particu-larly instructive. In its strict sense, Weber's law states, as is well known, that the size of "differential thresholds" (smallest per-ceptible differences) is proportional to that of the elements compared; for example, if a subject distinguishes 10 mm. from 11 mm. but not 10 from 10.5 mm., he will also only distinguish 10 from 11 cm. and not 10 from 10.5 cm.

Let us now assume that the aforementioned lines A and A' are of equal or nearly equal values. If they are equal, centring on A enlarges A and decreases A', and centring on A' enlarges A' and decreases A in the same proportions; hence the distor-tions are cancelled. On the other hand, if they are slightly unequal but with an inequality which is less than the distor-tions due to centralisation, then centring on A yields the per-ception A > A' and centring on A' the impression A' > A. In this case, there is a contradiction between the estimations (as opposed to the general case where an inequality, common to both methods of viewing, simply appears greater or smaller

[1] See note p. 67.

according to whether A or A′ is fixated). This contradiction is interpreted as a sort of fluctuation (comparable to resonance in physics) which can arrive at perceptual equilibrium only by the equation $A = A'$. But this equation remains subjective and is therefore illusory; it amounts to saying that two almost equal values are confused in perception. Now this non-differentiation is precisely what characterises the existence of "differential thresholds" and since, by the law of relative centralisation, it is proportional to the lengths of A and A′, we thus return to Weber's law.

Weber's law applied to differential thresholds is thus explained by the law of relative centralisation. Moreover, since it applies with equal force to differences of any description (whether the similarities exceed the differences, as in cases below threshold value, or whether the reverse is the case as in the case discussed above), we may in all cases regard it as simply expressing the factor of proportionality inherent in the relations between relative centrings (and for touch, weight, etc. just as for vision).

We are now in a position to state more clearly the undoubtedly essential opposition which separates intelligence from perception. Weber's law is often translated by saying that all perception is "relative". Absolute differences are not discerned since 1 gr. may be perceived when added to 1 gr. although it is not when it is added to 100 gr. On the other hand, when the elements differ markedly the contrasts are then accentuated, as is shown by ordinary cases of relative centralisation, and this reinforcement is again relative to the size of the values involved (thus a room seems warm or cold according as one comes from a place with a higher or lower temperature). Thus whether we are concerned with illusory similarities (threshold of equality) or illusory differences (contrasts), perceptually they are all "relative". But does not the same hold in the case of intelligence also? Is not a class relative to a classification and a relation to a complex of relations? In point of fact, the word "relative" is used in quite different senses in the two cases.

Perceptual relativity is a distorting relativity, in the sense in which conversational language says "everything is relative" when denying the possibility of objectivity; the perceptual relation modifies the elements which it unites, and we now understand why. The relativity of intelligence on the other hand is the very condition of objectivity; thus relativity in space and in time is a condition of their very measurement. Everything indicates, therefore, that perception, obliged to proceed step by step by immediate but partial contact with its object, distorts it by the very act of centring it, although these distortions are reduced by equally partial decentralisations, while intelligence, encompassing in a single whole a much larger number of facts reached by variable and flexible paths, attains objectivity by a much more thorough decentralisation.

These two relativities, the one distorting and the other objective, are doubtless the expression both of a deep-rooted opposition between acts of intelligence and perception, and of a continuity which, in other respects, presupposes the existence of common mechanisms. If perception, like intelligence, consists in structuring and arranging relations, why then are these relations distorting in one case and not in another? Might not the reason be that the first are not only incomplete but cannot be sufficiently co-ordinated, while the second are based on a co-ordination which can be indefinitely generalised? And if the "grouping" is the source of this co-ordination, and if its principle of reversible combinativity carries further the work of perceptual regulations and decentralisations, should we not then admit that centrings are distorting because they are not numerous enough, being to some extent fortuitous and so resulting from a sort of lottery among those which would be necessary to ensure complete decentralisation and objectivity?

We are therefore led to enquire whether the essential difference between intelligence and perception does not arise from the fact that the latter is a process of a statistical nature, confined to a certain stage, while processes of an intellectual nature determine complex relations confined to a higher level. Perception would be to intelligence what, in physics, irreversible functions (i.e. simple chance functions) and displacements of equilibrium are to mechanics proper.

The probabilist structure of the perceptual laws of which we have been speaking amounts precisely to the same as, and explains, the irreversible character of the processes of perception, as opposed to operational combinations, which are both well defined and reversible. Why does sensation appear as the logarithm of the excitation (which immediately explains the proportionality expressed by Weber's law)? It is known that Weber's law applies not only to facts of perception or facts of physiological excitation but also, among other things, to the impression on a photographic plate. In this last case it means simply that the intensity of the impression is a function of the probability of a collision between the photons bombarding the plate and the particles of silver salts which compose it (hence the logarithmic form of the law: the relation between the multiplication of probabilities and the addition of intensities). Similarly, in the case of perception, it is easy to think of a quantity such as the length of a line, as a mass of possible points of visual fixation (or of segments for possible centralisation). When two unequal lines are compared, the corresponding points will give rise to combinations or associations (in the mathematical sense) of similarity, and the non-corresponding points to associations of difference (the associations thus increasing geometrically as the length of the lines increases arithmetically). If perception occurred according to every possible combination, there would then be no distortion (the associations would reach a constant relation and we should have

$s = -d$). But the facts suggest that actual vision constitutes a sort of sampling, as though only certain points of the perceived figure were fixated while others were neglected. It is easy, then, to interpret the foregoing laws in terms of probabilities according to which centrings are more likely to be placed in one direction than another. In the case of a considerable difference between two lines, it follows automatically that the larger of the two will catch the eye more, hence the excess of associations of difference (the law of relative centralisation concerning contrast), while in the case of very small differences, associations of similarity will outweigh others, hence Weber's threshold.[1] (We may even calculate the various combinations and again arrive at the formulae stated above.)

Finally, we may note that this probabilist character of perceptual constructions, as opposed to the determinate character of operational combinations, not only explains the distorting relativity of the first and the objective relativity of the second. Above all, it explains the essential fact which Gestalt psychology has insisted on: namely, that in a perceptual structure the whole is not reducible to the sum of its parts. In fact, to the extent that chance is a factor in a system, that system will be prevented from being reversible, since this chance factor always, in one way or another, involves the existence of a mixture and a mixture is irreversible. The result is that a system involving an element of chance could not be liable to additive combination (inasmuch as reality overlooks extremely unlikely combinations), unlike determinate systems which are reversible and combinative.[2]

---

[1] See Piaget. 'Essai d'interprétation probabiliste de la loi de Weber." *Arch. de Psychol.* XXX (1944) pp. 95–138.

[2] The best example of non-additive combination of a perceptual type is doubtless provided by a certain weight illusion wherein the part A (a piece of cast iron) is perceived as heavier than the whole B, comprising A and A′ (an empty box of light wood exactly enclosing A). Thus B < A − A′, and A > B while objectively B = A + A′.

So, in short, we can say that perception differs from intelligence in that its structures are intransitive, irreversible, etc. and thus not composed in accordance with laws of grouping, the reason for this being that the distorting relativity which is inherent in them gives them an essentially statistical nature. This statistical composition of perceptual relations is thus simply the same as their irreversibility and their non-additivity, while intelligence is directed towards complete and therefore reversible combinativity.

ANALOGIES BETWEEN PERCEPTUAL ACTIVITY AND INTELLIGENCE

How then are we to explain the undeniable affinity between these two types of structure, both of which imply constructive activity on the part of the subject and constitute complex systems of relations, certain of which, in both fields, arrive at "constancies" or at notions of conservation ? Above all, how are we to account for the existence of the innumerable intermediate structures which link elementary centralisations and decentralisations, as well as the regulations resulting from the latter, with intellectual operations?

It seems that, in the perceptual field, a distinction must be made between perception as such—the totality of relations given immediately and simultaneously with each centring—and the perceptual activity which comes into play in the very act of centring vision or of changing the centring (as well as in other acts.) It is clear that this distinction is still relative, but it is remarkable that each school should be obliged to recognise it in one form or another. In this way, the Gestalt theory, whose whole character tends to restrict the subject's activity in favour of whole structures, which are prominent by virtue of both physical and physiological laws of equilibrium, has been forced to find a place for the subject's attitudes. The "analytical attitude" is invoked to explain how wholes may be partially dissociated and, especially, the *Einstellung* or mental set of the subject is admitted as the cause of numerous distortions in perception depending on previous states. As for von Weizsäcker's school, Auersperg and Buhrmester invoke anticipations and perceptual reconstitutions, which involve the necessary intrusion of the response in all perception. And so on.

Now if a perceptual structure is essentially of a statistical nature and not composed additively, it follows automatically that all activity which directs and co-ordinates successive centrings will reduce the role of chance and change the structure concerned in the direction of operational composition (needless to say, in varying degrees and without ever completely attaining it). Side by side, therefore, with the manifest differences between the two fields, there exist analogies, which are no less obvious and such that it would be difficult to say just where perceptual activity ends and intelligence begins. This is why nowadays we cannot speak of intelligence without defining its relations with perception.

The crucial fact, in this connection, is the existence of a perceptual development as a part of mental growth in general. Gestalt Psychology has rightly insisted on the relative invariability of certain perceptual structures; most illusions occur at all ages, and in the animal just as in man; factors determing complex "configurations" likewise appear to be common to all ages, etc. But these common mechanisms especially concern perception as such, which is in some way receptive and immediate,[1] while perceptual activity itself and its effects manifest far-reaching transformations, varying with mental age. As well as size constancy, etc., which experiment has shown, despite the Gestalt theory, to be built up gradually with the appearance of ever more precise regulations, the simple measurement of

---

[1] This does not mean "passive", since it already shows "laws of organisation".

illusions shows the existence of modifications with age that would be inexplicable without a close affinity between perception and intellectual activity in general.

Here we must distinguish two cases, corresponding on the whole to what Binet called "innate" and "acquired" illusions, and which we had best straight away name "primary" and "secondary" illusions. Primary illusions are reducible to simple factors of centralisation and are thus dependent on the law of relative centralisation. Now the value of these diminishes fairly regularly with age ("error of the standard", illusions of Delbœuf, Oppel, Müller-Lyer, etc.) and this is readily explained by the increase in decentralisations, and in the regulations which they involve, as the subject's activity when faced with the figures increases. Certainly, the young child remains passive where older children and adults compare, analyse and thus indulge in an active decentralisation which is orientated towards operational reversibility. But, on the other hand, there are illusions which increase in intensity with age or development, such as the weight illusion, which is absent in the grossly abnormal and which increases up to the end of childhood, to decrease somewhat afterwards. But we know that what it requires is simply a sort of anticipation of the relations of weight and volume, and it is clear that this anticipation presupposes an activity which by its very nature increases of its own accord with intellectual growth. Such an illusion, produced by interaction between primary perceptual factors and perceptual activity, may thus be called secondary and we shall shortly be meeting others which are of the same type.

This being so, perceptual activity is distinguished in the first place by the occurrence of decentralisation, which corrects the effects of centralisation and thus constitutes a regulation of perceptual distortions. Now, however elementary and dependent on sensori-motor functions these decentralisations and regulations may be, it is clear that they all constitute an activity of comparison and co-ordination which is allied to that of intelligence. Even to look at an object is an act and by noting whether a young child lets his gaze dwell on the first point that presents itself or whether he directs it so as to include the whole complex of relations, we can almost judge his mental age. When objects that are too distant to be included in the same centring are to be compared, perceptual activity is extended in the form of "transportations" in space, as though the view of one of the objects were being superimposed on the other. These transportations, which thus constitute the (potential) reconciliation of centrings, give place to genuine "comparisons" or double transportations which, by alternating, decentralise the distortions due to one-way transportation. Study of these transportations has drawn our attention to a distinct reduction of distortions with age,[1] that is to say, a distinct improvement in the estimation of size at a distance, and this is self-explanatory in view of the coefficient of true activity which occurs here.

Now, it is easy to show that these decentralisations and double transportations, together with the specific regulations which their different varieties involve, are responsible for the famous perceptual "constancies" of shape and size. It is most remarkable that we scarcely ever obtain absolute size constancy in the laboratory; the child under-estimates size at a distance (taking into account the error of the standard), but the adult almost always over-estimates it slightly! These "superconstancies", which writers have in fact often observed but which they normally pass over as though they were embarrassing exceptions, have seemed to us to constitute the rule, and no fact could better attest the intervention of true regulation in the construction of constancies. Now when we see that infants, just at the age at which this constancy has been noted (although its precision has been greatly exaggerated), indulge in genuine trials, which consist in

[1] *Arch. de Psychol.*, XXIX (1943) pp. 173–253.

deliberately moving to and from their eyes the objects they are looking at,[1] we are induced to relate perceptual activity involving transportations and comparisons to manifestations of sensori-motor intelligence (without resorting to Helmholtz's "unconscious inference"). On the other hand, it seems obvious that the shape constancy of objects is bound up with the actual construction of the object. We shall return to this in the next chapter.

In brief, perceptual "constancy," seems to be the product of genuine actions, which consist of actual or potential movements of the glance or of the organs concerned; responses are co-ordinated in systems, whose organisation may vary from simple directed trial-and-error to a structure reminiscent of "grouping". But true grouping is never attained at the perceptual level, and only regulations due to these real or potential movements take place. This is why perceptual "constancy", although it recalls operational constants or ideas of conservation depending on reversible and grouped operations, does not arrive at the ideal precision which alone would assure them the complete reversibility and mobility of intelligence. Nevertheless, the perceptual activity that characterises it is already approaching intellectual combinativity.

This same perceptual activity likewise presages intelligence in the domain of temporal transportations and genuine anticipations. In an interesting experiment on visual analogies of the weight illusion, Usnadze[2] gives his subjects a fraction of a second's glimpse of two circles, 20 and 28 mm. in diameter, and then two circles of 24 mm. The 24 mm. circle, situated in the place previously occupied by the 28 mm. circle, is then seen as smaller than the other (and the one replacing the 20 mm. one is overestimated), on account of a contrast effect due to

[1] *La Construction du Réel chez l'Enfant*, pp. 157–158.
[2] *Psychol. Forsch.*, XIV (1930), p. 366.

transportation in time (which Usnadze calls *Einstellung*), Measuring this illusion in children aged 5–7 and in adults,[1] we, with Lambercier's collaboration, obtained the following results, and it is very suggestive to consider them as a whole with regard to the relations of perception to intelligence. On the one hand, the Usnadze effect is appreciably stronger in adults than in children (as is the weight illusion itself), but, on the other hand, it disappears more rapidly. After several presentations of 24 + 24 mm. the adult reverts gradually to objective vision, while the child retains a residual effect. So we cannot explain this double difference in terms of simple memory traces without being compelled to say that the adult's memory is superior but he forgets more quickly! On the contrary, it looks as if an activity of transposition and anticipation develops with age, towards both greater mobility and greater reversibility. This constitutes a fresh example of perceptual development orientated towards the operation.

A neat experiment of Auersperg and Buhrmester consists in presenting a simple square, traced in white lines, which is rotated on a black disk. At slow speeds the square is seen directly, although the retinal image now consists of a double cross enclosed by four lines at right angles. At high speeds, only the retinal image is seen, but at intermediate speeds a transitional figure is seen, formed by a simple cross enclosed by the four lines. As these writers have emphasised, a sensori-motor anticipation undoubtedly occurs which enables the subject to reconstruct the square wholly (first phase), in part (second phase), or which miscarries (third phase), being upset by the excessive speed. Now, we have found, with Lambercier and Demetriades, that the second phase (simple cross), measured in children aged 5–12 years, appeared later and later (i.e. at higher and higher speeds of rotation as age

[1] *Arch. de Psychol.*, XXX (1944), pp. 139–196.

increased). The reconstruction or anticipation of the moving square thus improves (i.e. occurs at ever higher speeds) as the subject develops.

But this is not all. We present the subjects with two rods to be compared in depth perception, A at 1 metre and C at 4 metres. We first measure the perception of C (underestimation or super-constancy, etc.) then we place on this side of C a rod B equal to A, placed 50 cm., away from it laterally, or else we place between A and C an intermediate series B1, B2, B3, all equal to A (with the same lateral interval). The adult, or the child older than 8–9 years, immediately sees $A = B = C$ (or $A = B1 = B2 = B3 = C$) because he transports the perceptual equalities $A = B$ and $B = C$ directly to the relation $C = A$, thus closing the figure. Young children, on the other hand, see $A = B$, $B = C$ and A different from C, as though the equalities seen in the course of the detour A B C were not transferred to the direct relation A C. Now before 6–7 years the child is not capable of the operational combination of the transitive relations $A = B$, $B = C$, therefore $A = C$. But, curiously enough, between 7 and 8–9 years there is an inter-mediate phase when the subject immediately infers by intelli-gence the equality $A = C$ while at the same time he sees C per-ceptually as slightly different from A! It is clear then from this example that transposition (which is a "transportation"of rela-tions as opposed to that of an isolated value) also arises from perceptual activity and not from the automatic structuring common to all ages, and that we still have to define the relations between perceptual transposition and operational transitivity.

But transposition is not merely external to the perceived fig-ures; as well as this external transposition we must distinguish internal transpositions which enable us to recognise, within the actual figures, recurring relations, symmetries (or reversed rela-tions), etc. Here also, there is much to be said concerning the role of intellectual development, young children being by no means so apt at structuring complex figures as some people have tried to maintain.

From all these facts, we may conclude the following. The development of perception bears witness to the existence of a perceptual activity leading to decentralisations, transportations (spatial and temporal), comparisons, transpositions, anticipations and, in general, an analysis becoming more and more mobile and making for reversibility. This activity increases with age and it is because they do not possess it to a sufficient degree that young children perceive in a 'syncretic' or 'global' manner or else by accumulating disconnected details.

While perception as such is characterised by irreversible systems of a statistical nature, perceptual activity, on the other hand, introduces into such systems, which are governed by fortuitous or merely probable distributions of centrings, coherence and the power of progressive synthesis. Does this activity already constitute a form of intelligence? We have seen (Chap. 1 and end of Chap. 2) what little meaning a question of this type has. However, we can say that in their origins the actions that serve to co-ordinate attention along the lines of decentralisation, transportation, comparison, anticipation and especially transposition are closely bound up with sensori-motor intelligence, which we shall be discussing in the next chapter. Transposition, in particular, both internal and external, which embraces all other acts of a perceptual nature, is very much like assimilation, which is a characteristic of sensori-motor schemata, and especially like the generalised assimilation that facilitates the transference of these schemata.

But, if perceptual activity approaches sensori-motor intelligence, its development takes it up to the threshold of operations. In proportion as the perceptual regulations due to comparisons and transpositions tend towards reversibility, they constitute one of the flexible supports which will be required for the launching of the operational mechanism. The latter, once established, will then react on them, integrating them with itself by a recoil analogous to that occuring in the example we have just mentioned in connection with transpositions of equality. But, prior to this reaction, they pave the way for the operation, introducing more and more mobility into the sensori-motor mechanisms that constitute its substructure. In fact, it is sufficient that the activity underlying perception should pass beyond immediate contact with the object and act at increasing distances in space and time, for it to transcend the perceptual field itself and thus for it to be liberated from the limitations that prevent it from attaining complete mobility and reversibility.

However, perceptual activity is not the only medium of incubation provided for the generation of operations of intelligence; we still have to consider the role of the motor functions which produce habits and which are, moreover, extremely closely linked with perception itself.

4

HABIT AND SENSORI-MOTOR INTELLIGENCE

The distinction between motor functions and perceptual functions is legitimate only for purposes of analysis. As von Weizsäcker[1] has convincingly shown, the classical division of phenomena into sensory stimuli and motor responses, which is introduced by the reflex-arc schema, is just as fallacious, and refers to laboratory products which are just as artificial, as the idea of the reflex arc itself, conceived in isolation. Perception is influenced by motor activity from the outset, just as the latter is by the former. This is what we, for our part, have asserted when speaking of sensori-motor schemata in order to describe the simultaneously perceptual and motor assimilation which characterises the behaviour of the infant.[2]

We are bound then to place what we have just learned from the study of perception in its true genetic context and to inquire

[1] *Der Gestaltkreis*, 1941.
[2] *La naissance de l'intelligence chez l'enfant*, 1936.

how intelligence is formed prior to language. Once he has passed beyond the level of purely hereditary connections, i.e. reflexes, the infant acquires habits as a result of experience. Do these habits provide the basis for intelligence or have they nothing to do with it? This is the parallel problem to the one we put to ourselves with regard to perception. The answer also is likely to be the same, and this will enable us to advance more rapidly, and to place the development of sensori-motor intelligence in relation to all the elementary processes that condition it.

HABIT AND INTELLIGENCE

I. Independence or direct derivation

Nothing is better fitted to illustrate the continuity which links the problem of the birth of intelligence to that of the formation of habits than a comparison of the various answers to these two questions. The same hypotheses appear in both cases, as though intelligence were an extension of those mechanisms which in their automatic form appear as habit.

In connection with habit, we once again find the genetic schemata of association, of trial-and-error or of assimilatory structuring. In its treatment of the relations between habit and intelligence, associationism goes so far as to make habit into a primary fact which explains intelligence; the theory of trial-and-error reduces habit to a matter of responses selected in the course of random behaviour and becoming automatic, this being characteristic of intelligence itself; the theory of assimilation sees intelligence as a form of equilibrium of that assimilatory activity which, in its original form, constitutes habit. As for non-genetic interpretations, we shall again meet the three combinations corresponding to vitalism, apriorism and Gestalt: habit deriving from intelligence, habit unrelated to intelligence and habit explained, like intelligence and perception, by structurings whose laws remain independent of development.

Regarding the relations between habit and intelligence (the only question which concerns us here), we must ascertain first whether the two functions are independent, then whether the one derives from the other, and finally from which common forms of organisation they emanate at different levels.

It is typical of the logic of the apriorist interpretation of intellectual operations to deny that they have any relation to habits, since they emanate from an inner structure which is independent of experience. And it is a fact that, to an introspection of the two types of phenomena in their final state, their differences seem profound and their analogies superficial. H. Delacroix has shrewdly commented on both of them: a habitual response to repeated circumstances seems to involve a sort of generalisation, but, in place of this unconscious automatism, intelligence substitutes a generality of quite a different quality, composed of deliberate choices and insight. All this is quite correct, but the more we analyse the formation of a habit as opposed to the automatic exercising of it, the more we realise the complexity of the activities that come into play at the outset. On the other hand, in going back to the sensori-motor origins of intelligence, we come back to the setting of the learning process in general. So, before deciding that the two types of structure are irreducible, it is essential to inquire, while distinguishing vertically a series of actions of different levels, and while taking account horizontally of how far they are novel or automatic, whether there might not exist a certain continuity between the limited and comparatively rigid co-ordinations that we usually call habits, and the co-ordinations characterising intelligence, which have greater mobility and extreme limits which are further removed.

This was fully realised by Buytendijk, who has brilliantly analysed the formation of elementary animal habits, especially in invertebrates. However, the greater the complexity this writer finds in the factors affecting habit, the more he tends, on account of his vitalist system of interpretation, to subordinate the co-ordination peculiar to habits to intelligence itself, a faculty inherent in the organism as such. The formation of a habit always involves a fundamental means-end relation; an action is never a succession of mechanically associated movements but is directed towards a satisfaction such as contact with food or release, e.g. Limnaea, when placed upside down, return more and more rapidly to their normal position. But the means-end relation characterizes intelligent actions; habit would then be the expression of an intelligent organisation which, moreover, must be co-extensive with all living structure. Just as Helmholtz explained perception by the intervention of unconscious infer-ence, so vitalism ends by describing habit as the result of an unconscious organic intelligence.

But although we must fully acknowledge the justice of Buytendijk's observations regarding the complexity of the sim-plest acquisitions and the irreducibility of the relation of need to satisfaction, which is the origin and not the effect of associ-ations, there is no justification for hastily explaining everything by intelligence, considered as a primary fact. Such a thesis involves a series of difficulties which are exactly the same as those of the parallel interpretation with respect to perception. In the first place, habit, like perception, is irreversible because it is always orientated in one direction towards the same result, while intelligence is reversible. Reversing a habit (e.g. writing backwards or from right to left, etc.) means acquiring a new habit, while a "reverse operation" of intelligence is psychologic-ally implied by the original operation (and logically constitutes the same change, but in the opposite direction). In the second place, just as intelligent insight only slightly modifies a percep-tion (knowledge has little influence on an illusion as Hering pointed out in reply to Helmholtz) and, reciprocally, elementary perception does not automatically turn itself into an act of intelligence, so intelligence only slightly modifies an acquired habit and, above all, the formation of a habit is not immediately followed by the development of intelligence. There is actually an appreciable break in the genetic series between the appearance of the two types of structures. Piéron's sea-anemones, which close up at low tide and thus store the water they need, are not evidence for a really mobile intelligence and, in particular, they retain their habit in the aquarium for several days before it is extinguished. Goldsmith's Gobii learn to pass through a hole in a sheet of glass to reach food and keep to the same route after the glass is removed: we may name this behaviour sub-cortical intelligence, but it is still considerably inferior to what is ordinarily called intelligence without qualification.

Hence the hypothesis which for a long time seemed the simplest: that habit constitutes a primary fact, explicable in terms of passively experienced associations, and intelligence grows out of it gradually, by virtue of the growing complexity of the acquired associations. We are not going to call associationism to trial here, since the objections to this mode of interpretation are as well known as its resurrection in different and often disguised forms. However, it is essential, in order to arrive at the true development of the structures of intelligence, to remember that the most elementary habits are still irreducible to the pattern of passive association.

But the idea of the conditioned reflex, or of conditioning in general, has afforded a recrudescence of vitality to associationism by providing it with both a precise physiological model and a revised terminology. Hence the series of applications attempted by psychologists in the interpretation of intellectual functions (language etc.) and occasionally of the act of intelligence itself.

But if the existence of conditioned behaviour is a fact, and even a very important one, its interpretation does not imply the reflexological associationism with which it is too often identified. When a response is associated with a perception there is more in this connection than a passive association (i.e. becoming stamped in as a result of repetition alone); meanings also enter into it, since association occurs only in the presence of a need and its satisfaction. Everyone knows in practice, although we too often forget it in theory, that a conditioned reflex is stabilised only as long as it is confirmed or reinforced; a signal associated with food does not give rise to an enduring reaction if real food is not periodically presented together with the signal. Association thus comes to be part of a complex piece of behaviour, which starts from a need and finishes with its satisfaction (actual, anticipated or even make-believe, etc.). This amounts to saying that this is not a case of association in the classical sense of the term, but rather of the constitution of a complex schema bound up with a meaning. Moreover, if a system of conditioned responses is studied with reference to their historical sequence (and those concerning psychology always present such a sequence, as opposed to over-simplified physiological conditioning), the role of complex structuring is seen to even better advantage. Thus André Rey placed a guinea-pig in compartment A of a box with three adjacent compartments, A, B, and C, and administered an electric shock preceded by a signal. On the repetition of the signal, the guinea-pig jumped into B, then returned to A, but only a few more trials were required for it to jump from A into B, from B into C and to return from C into B, and so into A. Thus, in this case, the conditioned response is not the simple substitution of responses originally due to a simple reflex, but new behaviour which arrives at stability only by a structuring of the whole environment.[1]

Now if this is the case with the most elementary types of habit, the same must hold a fortiori in the case of the increasingly

[1] A. Rey, "Les conduites conditionnées du cobaye" (*Arch. de Psychol.* XXV (1936), pp. 217–312).

complex "associative transfers" which carry behaviour to the threshold of intelligence. Wherever there is an association between response and perception, the so-called association really consists in integrating the new element with a previous schema of activity. Whether this previous schema is in the nature of a reflex, as in the conditioned reflex, or belongs to much higher levels, association is always, in point of fact, assimilation of such a kind that the associative link is never simply the reproduction of a relation which is given, already formed, in external reality.

This is why the study of the formation of habits, like that of the structure of perceptions, concerns the problem of intelligence in the highest degree. If early intelligence consisted merely in exerting its action (which is a later acquisition belonging to a higher plane) on a completed world of associations and relations, corresponding term for term with relations written, once and for all, in the external environment, then this action would, in point of fact, be illusory. On the other hand, in so far as the organising assimilatory process, which eventually arrives at the operations peculiar to intelligence, appears from the outset in perceptual activity and in the formation of habits, the empiricist models of intelligence that some writers try to build up are inadequate at all levels, since they disregard assimilatory construction.

We know, for example, that Mach and Rignano regard reasoning as a "mental experiment". This description, correct in principle, would take the form of an explanatory solution if the experiment were the copy of a cut-and-dried external reality. But as this is not so and as, even at the level of habit, adaptation to reality means an assimilation of reality to the subject's schemata, the explanation of reasoning as a mental experiment becomes circular; the whole activity of intelligence is required to carry out an experiment, practical or mental. In its finished state, a mental experiment is the reproduction in thought, not of reality, but of actions or operations which affect it, and the problem of their formation remains untouched. Only at the level at which thought begins in the child may we speak of mental experiment in the sense of a simple internal imitation of reality; but in this case reasoning is, of course, not yet logical.

Similarly, when Spearman reduces intelligence to three essential activities, the "apprehension of experience", the "eduction of relations" and the "eduction of correlates", we must add that experience is not apprehended without the intervention of constructive assimilation. The so-called "eductions" of relations are to be thought of, then, as genuine operations (seriation or the grouping together of symmetrical relations). As for the eduction of correlates "the presenting of any character together with any relation tends to evoke immediately the knowing of the correlative character",[1] this is compatible with certain definite groupings, namely those of multiplication of classes or relations (Chap. 2).

HABIT AND INTELLIGENCE

II. Trial-and-error and structuring

So if neither habit nor intelligence may be explained by a system of associative co-ordinations that correspond exactly to relations previously given in external reality, but both instead involve action on the part of the subject himself, would not the simplest interpretation be to reduce this activity to a series of trials occurring at random (i.e. with no direct relation to the environment), but gradually selected by means of the successes or failures resulting from them? In this way, Thorndike, studying the mechanism of learning, places animals in a maze and measures retention by the decreasing number of errors. At first the animal

---

[1] *The Nature of Intelligence*, 1923, p. 91.

acts at random, i.e. it indulges in fortuitous trials, but errors are gradually eliminated and the successful trials retained, so determining subsequent routes. This principle of selection by the result obtained is called the "law of effect". The hypothesis is tempting: action on the part of the subject is introduced into the trials, and that of the environment into selection, and the law of effect allows for the role of needs and satisfactions which embrace all active behaviour.

Moreover, it is in the nature of an explanatory scheme of that sort to take account of the continuity which links the most elementary habits with the most highly developed intelligence. Claparède took up the concepts of trial-and-error and the subsequent empirical test and made them the basis of a theory of intelligence successively applied to animal intelligence, to the practical intelligence of the child, and even to the problem of "the genesis of the hypothesis"[1] in the psychology of adult thought. But in the numerous writings of this German psychologist we see a significant development from the first to the last, so that the mere study of this development constitutes, of itself, an adequate criticism of the concept of trial-and-error.

Claparède begins by opposing intelligence—a vicarious function for adaptation to new conditions—to (automatic) habit and instinct, which are adaptations to repeated circumstances. But how does the subject behave in the presence of new circumstances? From Jennings' infusoria to man (and the scientist himself when he is confronted with the unexpected), he acts by trial-and-error. This trial-and-error may be merely sensori-motor or it may be internalised in the form of "trials" in thought alone, but its function is always the same: to contrive solutions from which experience will select afterwards.

The complete act of intelligence thus involves three essential stages: the question which directs the quest, the hypothesis

[1] *Arch. de Psychol.*, XXIV (1933), pp. 1-155.

which anticipates solutions, and the process of testing which selects from them. However, two types of intelligence may be distinguished, one practical (or "empirical"), the other deliberate (or "systematic"). In the first the question appears in the form of a simple need, the hypothesis as a sensori-motor random trial, and the testing process as a mere series of failures or successes. In the second, the need is reflected in the question, trial-and-error is internalised as a search for hypotheses and the testing process anticipates the sanction of experience by means of an "awareness of relations", which is sufficient to discard false hypotheses and to retain true ones.

Such was the outline of the theory when Claparède approached the problem of the genesis of the hypothesis in the psychology of thought. Now, while emphasising the role that trial-and-error obviously retains in the most evolved forms of thought, Claparède was led, through his method of "thinking aloud", to locate it no longer at the actual point of departure of intelligent enquiry but, so to speak, in the margin, or in the vanguard, and only when the material exceeds the subject's understanding. The starting-point seems to him, on the other hand, to be provided by an attitude, the importance of which he had not hitherto stressed: once the enquiry has been directed by the need or the question (through a mechanism which in other respects is still considered mysterious), the first thing to occur in the presence of the data of the problem is an awareness of a system of simple "implicative" relations. These implications may be true or false. If true they are left untouched by experience. If false they are contradicted by experience, and only then does trial-and-error start. Thus the latter occurs only as a surrogate or supplement, i.e. as behaviour derived indirectly from the initial implications. Claparède concludes that trial-and-error is never pure; it is partly directed by the question and the implications, and it becomes really fortuitous only when the data outstrip these anticipatory schemata.

In what does "implication" consist? This is where the doctrine finds its widest scope and again links up with the problem of habit just as much as with that of intelligence itself. "Implication" is in essence almost the old "association" of the classical psychologists, but it is accompanied by a feeling of necessity arising from within and no longer from without. It is the manifestation of a "primitive tendency" without which the subject could not profit by experience at any level (p. 104). It is not due to the "repetition of a pair of elements", but, on the contrary, it is the source of the repetition of like material and "comes into being as soon as the two elements of the pair first meet". (p. 105). Thus experience can only refute or confirm it but it does not create it. But when experience imposes a coupling, the subject reinforces it with an implication. In fact, its roots would be found in William James' "law of coalescence" (the very law with which James explained association!): "The law of coalescence engenders implication at the level of action and syncretism at the level of representation" (p. 105). Claparède thus goes so far as to interpret the conditioned reflex in terms of implication; Pavlov's dog salivates at the sound of a bell, after having heard it at the same time as he saw food, because then the bell "implies" food.

This gradual reversal of the trial-and-error theory is worth careful examination. To begin with an apparently secondary point, would it not perhaps be a pseudo-problem to ask ourselves how the question or the need directs the search as though they existed independently of this search? The question and the need itself are, in fact, the expression of previously constructed mechanisms which are simply in a momentary state of disequilibrium. The need to suck presupposes the complete organisation of the sucking apparatus and, at the other extreme, questions such as "what?", "where?", etc., are the expression of classifications, spatial structures, etc., which are already wholly or partly constructed (see Chap. 2). It follows that the schema that directs the search is the one whose previous existence is necessary to explain the appearance of the need or the question; these mark the awareness of the quest and, like the quest, amount to a single act of assimilation of reality to this schema.

This being so, is it legitimate to regard implication as a primary fact which is both sensori-motor and intellectual, and the source of habit as well as of insight? It is to be understood, of course, that this term is not taken in its logical sense as a necessary link between judgments, but in the very general sense of any relation of necessity. Now do two elements, seen together for the first time, give rise to such a relation? To take one of Claparède's examples, does a black cat, seen by an infant, involve immediately from its first perception the relation "cat implies black?" If the two elements are really seen for the first time, with neither analogy nor anticipations, they are certainly already grouped together in a perceptual whole, in a Gestalt, which expresses in another form James' law of coalescence or the syncretism invoked by Claparède. It is clear enough that we are concerned here with more than an association, in so far as the whole results, not from the conjunction of the two elements originally seen separately, but rather from their immediate fusion through complex structuring. However, this is not a necessary link; it is the beginning of a possible schema which, however, will only engender relations felt to be necessary, as long as they form a genuine schema through transposition or generalisation (i.e. an application to new elements), in short, by introducing an assimilation. The assimilation, then, is the source of what Claparède calls implication. To speak schematically, the subject will not arrive at the relation "A implies x" on first perceiving an A with the quality x, but he will be led to the relation "A implies x" inasmuch as he assimilates A_2 to the schema (A), this schema being created precisely by the assimilation $A_2 = A$. The dog that salivates at the sight of food will not salivate in this way at the sound of a bell unless he assimilates it, as a sign or a part of the total act, to the schema of this action.

Claparède has good reason to say that repetition does not engender implication but that it appears only in the course of repetition, since implication is the internal product of the assimilation that ensures the repetition of the external act.

Now this necessary intervention of assimilation further supports the reservations that Claparède was himself induced to make with regard to the general role of trial-and-error. Firstly, it is obvious that when trial-and-error occurs it cannot be explained in mechanical terms. Mechanically, that is to say with the hypothesis of simple traces, error should be reproduced as often as successful trials. If such is not the case, i.e. if the "law of effect" holds, it means that the subject anticipates his failures and successes. In other words, each trial operates on the next, not as a channel opening the way to new responses, but as a schema enabling meanings to be attributed to subsequent trials.[1] So trial-and-error in no way excludes assimilation.

But that is not all. The very first trials are difficult to reduce to simple chance.[2] In maze experiments D. K. Adams finds responses directed from the outset. W. Dennis and also J. Dashiell lay stress on the continuation of the sets initially adopted. Tolman and Krechevsky even speak of "hypotheses" in describing the behaviour of rats, etc. Hence the important interpretations reached by C. L. Hull and E. C. Tolman. Hull insists on the contrast between psychological models involving means and ends and mechanistic models of path-tracing: while a straight line is the only possibility in the latter case, the former provide a number of possible paths which will be more numerous as the act is more complex. This amounts to saying that, from the level of sensori-motor behaviour onwards, which is intermediate between learning and intelligence, account must be taken of

[1] See *La naissance de l'intelligence chez l'enfant*, Chap. V and Guillaume, *La formation des habitudes*, pp. 144–154.

[2] *La formation des habitudes*, pp. 65–67.

what, in their final "groupings", becomes "the associativity" of operations (Chap. 2). As for Tolman, he brings out the role of generalization in the formation of habits themselves. Thus, when an animal is placed in a new maze different from the one known to it, it perceives general analogies and applies to the new case behaviour that met with success in the previous case (particular routes). So there is always complex structuring, but, for Tolman, the structures concerned are not simple "configurations" in the sense of Köhler's theory; they are *sign-gestalts*, i.e. schemata provided with meanings. This double property of general validity and meaning belonging to the structures considered by Tolman is a fairly good indication that he is concerned with what we call assimilatory schemata. Thus, from elementary learning to intelligence, there seems to be involved an assimilatory activity, which is as necessary to the structuring of the most passive forms of habits (conditioned responses and associative transfers) as it is to the unfolding of visible manifestations of activity (directed trial-and-error). In this respect, the problem of the relations between habit and intelligence is a fair parallel to that of the relations between intelligence and perception. Just as perceptual activity is not identical with intelligence, but links up with it as soon as it is freed from centring on the immediate and present object, so the assimilatory activity that engenders habits is not the same as intelligence but leads to the latter as soon as irreversible and isolated sensori-motor systems are differentiated and co-ordinated in mobile articulations. Besides this, the affinity between these two kinds of activity is obvious, since perceptions and habitual responses are constantly united in complex schemata, and since the "transfer" or generalisation characteristic of habit is the exact equivalent, on the motor side, of "transposition" in the domain of spatial figures, both involving the same generalized assimilation.

SENSORI-MOTOR ASSIMILATION AND THE BIRTH OF INTELLIGENCE IN THE CHILD

To explain how intelligence springs from the assimilatory activity which, at an earlier stage, engenders habits, is to show how, from the point at which mental life is dissociated from organic life, this sensori-motor assimilation is converted into ever more mobile structures which have an ever wider scope.

From hereditary structures onwards, we see, side by side with the internal and physiological organisation of reflexes, cumulative effects of practice and the beginnings of problem-solving, which mark the first reactions at a distance in space and time by which we defined "behaviour" (Chap. 1). A neonate who is spoon-fed will later have difficulty in feeding at the breast. When he is allowed to suck from the outset, his skill improves steadily; when placed at the breast, he finds the best position and will find it more and more rapidly. Although he sucks anything, he will soon reject a finger but retain the breast. Between feeds he will suck without food, and so on. These commonplace observations show that even within the closed field of hereditarily governed mechanisms there emerge the beginnings of reproductive assimilation of a functional order (practice), generalized or transpositive assimilation (extension of the reflex-pattern to new objects) and assimilation by recognition (discrimination between situations).

It is in this already active context that the first acquisitions due to experience come to find a place (since reflex action does not yet lead to any genuinely new acquisition but simply to consolidation). Whether we are concerned with an apparently passive co-ordination such as conditioning (e.g. a signal releasing a preparatory set for sucking), or with a spontaneous extension of the scope of reflexes (e.g. systematic thumb-sucking by co-ordination of the movements of the arm and the hand with those of the mouth), in both cases the elementary forms of the habit grow out of an assimilation of new elements to previous schemata which are in essence reflex-schemata. But it is important to realize that the extension of the reflex-schemata, through the incorporation of a new element, involves by this very fact the formation of a schema of a higher order (a genuine habit), which then integrates the lower schema with itself. So the assimilation of a new element to a previous schema implies the integration of the latter, in its turn, with a higher schema.

However, it goes without saying that at the level of these primary habits we cannot yet speak of intelligence. Compared with reflexes, habit has a greater range in space and time. But even when extended, these primary schemata are still separate and have no internal mobility or co-ordination among themselves. The generalizations of which they are capable are still merely motor transfers comparable with the simplest perceptual transpositions, and in spite of their functional continuity with later stages, there is still no reason to compare them in their structure with intelligence itself.

With a third level, however, which begins with the co-ordination of vision and prehension (between 3 and 6 months, usually about $4\frac{1}{2}$, new behaviour appears which represents a transition between simple habit and intelligence. Let us imagine an infant in a cradle with a raised cover from which hang a whole series of rattles and a loose string. The child grasps this and so shakes the whole arrangement without expecting to do so or understanding any of the detailed spatial or causal relations. Surprised by the result, he reaches for the string and carries out the whole sequence several times over. J. M. Baldwin called this active reproduction of a result at first obtained by chance a "circular reaction". The circular reaction is thus a typical example of reproductive assimilation. The first movement executed and followed by its result constitutes a complete action, which creates a new need once the objects to which it relates have returned to their initial stage; these are then assimilated to the previous action (thereby promoted to the status of a schema) which stimulates its reproduction, and so on. Now this mechanism is identical with that which is already present at the source of elementary habits except that, in their case, the circular reaction affects the body itself (so we will give the name "primary circular reaction" to that of the early level, such as the schema of thumb-sucking), whereas thenceforward, thanks to prehension, it is applied to external objects (we will call this behaviour affecting objects the "secondary circular reaction," although we must remember that these are not yet by any means conceived as substances by the child).

The secondary circular reaction, then, occurs in an early form in the structures characteristic of simple habits. As these are independent items of behaviour, which are repeated as wholes without any pre-established goal and affected by chance circumstances occurring during the process, they have in fact little in common with a complete act of intelligence, and we should beware of projecting into the subject's mind distinctions that we may make on his behalf between an original means (pulling the string) and a final goal (shaking the cradle cover), as well as of attributing to him conceptions of objects and space that we associate with a situation which for him is unanalysed and global. Nevertheless, as soon as the response has been reproduced several times, we see that it shows a double tendency towards disarticulation and the internal re-articulation of its elements and towards generalization or active transposition when presented with new material not directly related to previous material. Concerning the first point, it may be shown that after the child has followed the events in the order—string, shaking, rattles—the response becomes capable of rudimentary analysis; the sight of motionless rattles, and especially the discovery of a new object which has just been suspended from the cover, comes to stimulate reaching for the string. Although there is still no genuine reversibility present, it is clear that there is an increased mobility, and that there is almost an articulation of the response-pattern into a means (reconstructed afterwards) and an end (adopted afterwards). On the other hand, if the child is confronted with a completely new situation, such as the sight of something moving 2–3 yards from him, or even some sound in the room, he responds by seeking and pulling the same string, as though he were trying to restart the interrupted spectacle by "remote control". Now this new action (which clearly confirms the absence of any spatial contacts or understanding of causality) surely constitutes an early form of true generalization. Internal articulation, as well as this external transposition of the circular schema, heralds the imminent appearance of intelligence.

With a fourth stage comes greater precision: After 8–10 months the schemata constructed by secondary reaction during the previous stage become susceptible of co-ordination among themselves, some serving as means and others setting a goal for action. Thus, in order to grasp an objective placed behind a screen which either wholly or partly conceals it, the child will first remove the screen (so utilising the schemata of grasping or striking, etc.) and then seize the objective. Consequently, the goal is thereafter decided on before the means, since the subject has the intention of grasping the objective before he has that of removing the obstacle, which implies a mobile articulation of the elemental schemata composing the complex schema. Moreover, the new complex schema is susceptible of much greater generalization than previously. This mobility, coupled with an increase in generalization, is especially marked in the fact that the child, when confronted with a new object, tries his most recently acquired schemata in turn (grasping, striking, shaking, rubbing, etc.), so that these serve as sensori-motor concepts, so to speak, as though the subject were trying to understand the new object through its use (in the manner of "definitions by use" which recur much later at the verbal level).

Behaviour of this fourth level thus shows a twofold progress in the directions of mobility and of an extension of the scope of its schemata. The routes between the subject and the object followed by action, and also by sensori-motor reconstitutions and anticipations, are no longer direct and simple pathways as at the previous stages: rectilinear as in perception, or stereotyped and uni-directional as in circular reactions. The routes begin to vary and the utilisation of earlier schemata begins to extend further in time. This is characteristic of the connection between means and ends, which henceforth are differentiated, and this is why we may begin to speak of true intelligence. But, apart from the continuity that links it with earlier behaviour, we should note the limitations of this early intelligence: there are no inventions or discoveries of new means, but simply application of known means to unforeseen circumstances.

Two acquisitions characterise the next stage, both relating to the utilisation of past experience. The assimilatory schemata so far described are of course continually accommodated to external data. But this accommodation is, so to speak, suffered rather than sought; the subject acts according to his needs and this action either harmonizes with reality or encounters resistances which it tries to overcome. Innovations which arise fortuitously are either neglected or else assimilated to previous schemata and reproduced by circular reaction. However, a time comes when the innovation has an interest of its own, and this certainly implies a sufficient stock of schemata for comparisons to be possible and for the new fact to be sufficiently like the known one to be interesting and sufficiently different to avoid satiation. Circular reaction, then, will consist of a reproduction of the new phenomenon, but with variations and active experimentation that are intended precisely to extract from it its new possibilities. Thus, having discovered the trajectory of a falling object, the child tries to drop it in different ways or from different positions. This reproductive assimilation with differentiated and intentional accommodation may be called the "tertiary circular reaction".

Thenceforward, when schemata are co-ordinated with one another as means and ends, the child is no longer limited to applying known means to new situations; he differentiates the schemata serving as means by a sort of tertiary circular reaction and comes in consequence to discover new means. In this way, a series of responses grows up which everybody admits as having the character of intelligence, e.g. drawing an objective towards oneself by means of the base on which it rests, by means of a piece of string attached to it, or even by means of a stick used as an independent intermediary. But, however complex this latter behaviour may be, it is as well to realise that it does not arise all of a sudden but is prepared by a whole succession of relations and meanings due to the activity of previous schemata—the relation of means to end, the idea that one object may set another in motion, etc. In this respect, behaviour directed towards the base supporting the objective is the simplest; being unable to reach the objective, the subject grasps at the intervening objects (the cloth on which the desired toy is placed, etc.). The movements imparted to the objective by the grasping of the cloth are still without meaning at earlier levels; when in possession of the necessary relations, however, the subject is aware of the possible utilisation of the supporting base straight away. In such cases we see the true role of trial-and-error in the act of intelligence. As well as being directed both by the schema which assigns a goal to action, and by the schema selected as an initial means, trial-and-error is also ceaselessly directed during successive trials by schemata capable of giving a meaning to fortuitous events, which are thus intelligently utilised. Trial-and-error, then, is never pure, but only constitutes the process of active accommodation which works hand in hand with the assimilatory co-ordinations constituting the essence of intelligence.

Finally, a sixth stage, which occupies part of the second year, marks the completion of sensori-motor intelligence. Instead of new means being exclusively discovered by active experimentation, as at the previous level, there may henceforth be inventions by rapid internal co-ordination of processes now unknown to the subject. To this last category belong the phenomena of sudden restructuring described by Köhler in chimpanzees and Bühler's *Aha-Erlebnis* or experience of sudden insight. Thus, in children who have no occasion to experiment with sticks before the age of one year 6 months, the first contact with a stick affords insight into its possible relations with the objective to be reached, and this without actual trial-and-error. Similarly, it seems obvious that certain of Köhler's subjects discovered the use of the stick, so to speak, by looking and without previous practice.

The main problem, then, is to understand the mechanism of these internal co-ordinations, which imply both invention without trial-and-error and a mental anticipation closely related to representation. We have already seen how the Gestalt theory explains things by a simple perceptual restructuring without reference to past experience. But it is impossible not to see in the behaviour of an infant at this sixth stage the end-result of all the development characterizing the previous five levels. In fact, it is clear that once he becomes used to tertiary circular reactions and to the intelligent trial-and-error that constitutes true active experimentation the child sooner or later becomes capable of internalizing this behaviour. When the subject no longer acts when confronted with the data of a problem, and appears to be thinking instead (one of our children, after having tried without success to widen the opening of a box of matches by random behaviour, interrupted his activity, looked carefully at the chink then visible, then opened and closed his own mouth), everything seems to indicate that he continues his attempts, but with implicit trials or internalised actions (the imitative movements of the mouth in the foregoing example are a very clear indication of this sort of motor thinking). What happens then, and how do we explain the discovery that yields the sudden solution? Sensori-motor schemata that have become sufficiently mobile and amenable to co-ordination among themselves give rise to mutual assimilations, spontaneous enough for there to be no further need for actual trial-and-error and rapid enough to give an impression of immediate restructuring. Internal co-ordination of schemata will, then, bear the same relation to the external co-ordination of the earlier levels, as inner speech, a simple rapid, internalised rough draft of overt language, bears to outer speech.

But does the greater spontaneity and speed of assimilatory co-ordination between schemata fully explain the internalisation of behaviour, or does representation begin at the present level, thus indicating the transition from sensori-motor intelligence to genuine thought? Independently of the advent of language, which the child begins to acquire at this age (but which is absent in chimpanzees who are, nevertheless, capable of remarkably intelligent inventions), two types of behaviour at this sixth stage testify to the beginnings of representation, but beginnings which scarcely go beyond the rather rudimentary representation of chimpanzees. On the one hand, the child becomes capable of delayed imitation, i.e. of producing a copy which occurs for the first time after the perception of the model has disappeared; now whether delayed imitation is derived from imaginal representation or whether it causes it, it is certainly closely linked with it (we shall reconsider this problem in Chap. 5). On the other hand, the child simultaneously arrives at the simplest form of symbolic play, consisting in using the body to produce an action foreign to the present context (e.g. pretending to sleep for fun, while he is actually wide awake). Here again there appears a sort of image which is enacted, and therefore motor, but it is already almost representative. Do not these enacted images, character-istic of delayed imitation and of the early make-believe symbol, act as significants in the internalised co-ordination of schemata? This is what seems to be illustrated in the example we mentioned a short while ago of the child who used his mouth to imitate the widening of the visible gap in a box he was trying to open.

THE CONSTRUCTION OF THE OBJECT AND OF SPATIAL RELATIONS

We have just noted the remarkable functional continuity which links the successive structures built up by the child from the formation of elementary habits to the spontaneous and sudden acts of invention which characterize the highest forms of sensori-motor intelligence. The affinity between habit and intelligence thus becomes manifest, both arising, although at different levels, from sensori-motor assimilation. We must now reconsider what we said above (Chap. 3), concerning the affinity between intelligence and perceptual activity, both of which depend on sensori-motor assimilation at different levels; in the one this assimilation engenders perceptual transposition (a close relative of the transfer of habitual movements), and the other is characterized by specifically intelligent generalization.

Nothing is better fitted to illustrate the bonds between perception, habit and intelligence, which are so simple in their common origin and so complex in their manifold differentiations, than an analysis of the sensori-motor construction of the fundamental schemata formed by the object and by space (which, incidentally, are indissociable from causality and time). Actually, this construction is closely correlated with the development of the pre-verbal intelligence which we have just been considering. But it also requires a high degree of organization of perceptual structures and of completely integrated motor structures built up of habits.

What in fact is the schema of the object? In one essential

respect it is a schema belonging to intelligence. To have the concept of an object is to attribute the perceived figure to a substantial basis, so that the figure and the substance that it thus indicates continue to exist outside the perceptual field. The permanence of the object seen from this viewpoint is not only a product of intelligence, but constitutes the very first of those fundamental ideas of conservation which we shall see developing within the thought process (Chap. 5). But the very fact that it conserves itself and is even reducible to this conservation means that the solid object (the only sort to be considered in the first instance) also conserves its dimensions and its shape; shape and size constancy is a schema arising at least as much from perception as from intelligence. Finally, it goes without saying that both in perceptual constancy and in the conservation that goes beyond the frontiers of the present perceptual field the object is linked with a series of motor habits which are both the source and the effects of the construction of this schema. We thus see how much light it is bound to throw on the true relations between intelligence, perception and habit.

But how is the schema of the object constructed? At the reflex level there are certainly no objects, the reflex being a response to a situation, and neither the stimulus nor the action elicited involve anything more than the qualities attached to perceptual displays without any necessary substantial substrate. When the infant seeks and finds the breast it is not necessary for him to regard it as an object, and the conditions of sucking, together with the permanence of the relevant postures, are sufficient to account for his behaviour without the intervention of schemata. At the level of the earliest habits, recognition does not imply an object either, because recognition of a perceptual display does not imply any belief in the existence of the perceived element apart from present perceptions and recognitions; similarly, calling an absent person by crying merely requires an anticipation of his possible return as a familiar perceived figure, and not spatial localization of this person as a substantial object in an organized reality. On the other hand, to follow a moving figure with the eyes and to continue to look for it when it disappears, or to turn the head to look in the direction of a sound, etc., constitute the beginnings of a practical permanence, beginnings which are, however, closely tied to the action in progress. They are perceptuo-motor anticipations and expectancies, determined by immediately previous perception and response, and are not yet by any means active searches, distinct from the response already initiated or determined by present perception.

The fact that during the third stage (secondary circular reactions), the child becomes capable of grasping what he sees, allows us to verify these interpretations. According to C. Bühler, the subject at this stage succeeds in removing a cloth covering his face. But we have been able to show that at this same stage the child makes no attempt to remove a cloth placed over a desired object, and this is the case even when he has already initiated a movement of prehension towards the object when it was visible. He thus behaves as though the object were absorbed by the cloth and ceased to exist at the very moment that it left the perceptual field or else, and this amounts to the same thing, he possesses no behaviour enabling him to search for the object which has disappeared—whether by action (lifting the screen) or by thought (imagining). However, he is more likely at this level than at the previous one to attribute a sort of practical permanence or momentary continuation to the objective of an action in progress, e.g. returning to a toy after having been distracted (delayed circular reaction), anticipating the position of a falling object, etc. But it is the action that confers a momentary conservation on the object, and the object loses this after the action in progress has ceased.

On the other hand, at the fourth stage (co-ordination of familiar schemata) the child begins to seek for the object behind a screen; this constitutes the beginning of behaviour concerning specifically the hidden object, and consequently the beginning of the conservation of substance. But we often observe an interesting reaction which shows that this early substantiality is not yet individualized and consequently remains tied to action itself: if the child is looking for an object at A (e.g. under a cushion situated to his right) and the same object is moved, in his sight, to B (another cushion to his left), he first returns to A, as though the object which disappeared under B was to be found in its original position! In other words, the object is still involved in a total situation characterized by the action that has just led to success, and does not always entail individualization of substance or co-ordination of successive responses.

At the first stage, these limitations disappear, except in the case where representation of invisible paths is necessary for the solution of the problem, and at the sixth stage even this condition ceases to be a hindrance.

It is therefore evident that the conservation of the object is prepared by the continuation of habitual responses and is the product of a co-ordination of the schemata constituting sensorimotor intelligence. So the object, at first an extension of the co-ordinations typical of habit, is constructed by intelligence itself, and constitutes the first constant of intelligence—a constant which is necessary for the formation of space, of causality in space and, in general, for all forms of assimilation which transcend the present perceptual field.

But, if its connections with habit and intelligence are obvious, the relations of the object to the perceptual constancies of shape and size are no less so. At the third of the levels that we have distinguished, a child, presented with his feeding-bottle the wrong way round, tries to suck the glass bottom if he does not see the rubber teat at the other end. If he sees it, he turns the bottle round (proof that there is no motor disability). But if, after having sucked at the wrong end, he sees the whole of the bottle (i.e. presented to him vertically) and then watches it being turned round, even then he will not succeed in turning it once the teat has again become invisible; thus the teat seems to him to be absorbed by the glass, except when he can see it. This behaviour, which is typical of the non-conservation of the object, thus involves non-conservation of the actual parts of the bottle, that is to say non-conservation of shape. At the next stage, however, corresponding with the construction of the permanent object, the bottle is reversed at once, and is thus perceived as a shape which remains constant in its entirety, in spite of being rotated. At this same level, we see the child slowly moving his head and taking an interest in the changes of shape in an object due to perspective.

As for size constancy (whose absence during the early months has recently been demonstrated by Brunswik), it also is developed during the fourth and especially during the fifth stage. Thus, one often sees the infant moving an object that he is holding in his hand towards and away from his eyes, as though he were studying changes in size with distance. There is then a correlation between the development of these perceptual constancies and the intelligent conservation of the object.

Now it is easy to understand the connection between these two kinds of reality. If constancies are actually the product of transportations, transpositions and their regulations, it is clear that these regulative mechanisms come from motor functions as much as from perception. Perceptual constancy of shape and size would thus be guaranteed by a sensori-motor assimilation which "transports" or transposes the relations concerned when modifications of the position or distance of the perceived object occur; in the same way, the schema of the permanent object would be due to a similar sensori-motor assimilation, which induces a search for the object once it leaves the perceptual field, thus endowing it with a conservation that is derived from the extension of the subject's own actions, projected as a property of the external world. We may thus grant that the same assimilatory

schemata both govern the shape and size constancy of the perceived object (by "transportations" and transpositions) and
elicit a search for it when it is no longer perceived; thus, when it
disappears, the object is sought because it is perceived as constant and, when it reappears, it is perceived as constant because it
gives rise to active seeking when it is no longer perceived. The
two aspects of perceptual activity and intelligence are in fact
much less differentiated at the sensori-motor level than is the
case with perception and reflective intelligence, since the latter
depends on symbols consisting of words or images, while
sensori-motor intelligence depends only on perceptions
themselves and on responses.

We may thus regard perceptual activity, in general as well as in
the particular case of constancy, as an aspect of sensori-motor
intelligence—an aspect which is limited to the case of an object
entering into direct and immediate relations with the subject,
whereas sensori-motor intelligence goes beyond the perceptual
field, anticipating relations which are to be perceived subsequently and reconstructing those which have been perceived
previously. The unity of the mechanisms affecting sensori-
motor assimilation is thus complete, which incidentally is what
the Gestalt theory has had the merit of showing, but it must be
interpreted in terms of the activity of the subject, and thus of
assimilation, not in terms of static configurations imposed
independently of mental development.

But there now arises a problem whose discussion leads to the
study of space. Perceptual constancy is the product of simple
regulations and we saw (Chap. 3) that the absence at all ages of
absolute constancy and the existence of adult "superconstancy"
provide evidence for the regulative rather than operational character of the system. There is, therefore, all the more reason why
it should be true of the first two years. Does not the construction
of space, on the other hand, lead quite rapidly to a grouping
structure and even a group structure in accordance with

Poincaré's famous hypothesis concerning the psychologically primary influence of the "group of displacements?"

The genesis of space in sensori-motor intelligence is completely dominated by the progressive organisation of responses, and this in effect leads to a "group" structure. But, contrary to Poincaré's belief in the *a priori* nature of the group of displacements, this is developed gradually as the ultimate form of equilibrium reached by this motor organisation. Successive co-ordinations (combinativity), reversals (reversibility), detours (associativity) and conservations of position (identity) gradually give rise to the group, which serves as a necessary equilibrium for actions.

At the first two stages (reflexes and elementary habits), we could not even speak of a space common to the various perceptual modalities, since there are as many spaces, all mutually heterogeneous, as there are qualitatively distinct fields (mouth, visual, tactile, etc.). It is only in the course of the third stage that the mutual assimilation of these various spaces becomes systematic owing to the co-ordination of vision with prehension. Now, step by step with these co-ordinations, we see growing up elementary spatial systems which already presage the form of composition characteristic of the group. Thus, in the case of interrupted circular reaction, the subject returns to the starting-point to begin again; when his eyes are following a moving object that is travelling too fast for continuous vision (falling etc.), the subject occasionally catches up with the object by displacements of his own body to correct for those of the external moving object.

But it is as well to realise that, if we take the point of view of the subject and not merely that of a mathematical observer, the construction of a group structure implies at least two conditions: the concept of an object and the decentralisation of movements by correcting for, and even reversing, their initial egocentricity. In fact, it is clear that the reversibility characteristic of the group presupposes the concept of an object, and also vice versa, since to retrieve an object is to make it possible for oneself to return (by displacing either the object itself or one's own body). The object is simply the constant due to the reversible composition of the group. Furthermore, as Poincaré himself has clearly shown, the idea of displacement as such implies the possibility of differentiating between irreversible changes of state and those changes of position that are characterized precisely by their reversibility (or by their possible correction through movements of one's own body). It is obvious, therefore, that without conservation of objects there could not be any "group", since then everything would appear as a "change of state". The object and the group of displacements are thus indissociable, the one constituting the static aspect and the other the dynamic aspect of the same reality. But this is not all: a world with no objects is a universe with no systematic differentiation between subjective and external realities, a world that is consequently "adualistic" (J. M. Baldwin). By this very fact, such a universe would be centred on one's own actions, the subject being all the more dominated by this egocentric point of view because he remains un-self-conscious. But the group implies just the opposite attitude: a complete decentralisation, such that one's own body is located as one element among others in a system of displacements enabling one to distinguish between one's own movements and those of objects.

This being so, it is clear that throughout the first two stages, and even in the third, none of these conditions is fulfilled; the object is not constituted and the different spaces, and later the single space that tends to co-ordinate them, remain centred on the subject. From then on, even when there seems to be (in practice) a return and a co-ordination in the form of a group, it is not difficult to distinguish appearance from reality, the latter constantly testifying to a privileged centralisation. In this way, an infant at the third stage who sees an object move along the line

AB to pass behind a screen at B, does not look for it at C at the other end of the screen, but looks back towards A; and so on. The moving object is therefore not yet an independent "object" following a rectilinear trajectory that is dissociated from the subject, but remains dependent on the preferred position A where it was first seen by the subject. As far as rotation is concerned, we noted above the example of the reversed feeding-bottle which is sucked at the wrong end instead of being turned round; this again attests the primacy of an egocentric perspective and the absence of the concept of an object, all of which explains the absence of any "group".

With the search for an object that has disappeared behind a screen begins the attribution of objectivity to co-ordinations and thus the construction of the sensori-motor group. But the very fact that the subject does not take account of successive displacements of the object and looks for it behind the first of the screens (see above), shows clearly enough that this nascent group is still partly "subjective", i.e. centred on the subject's own action, since the object remains dependent on the latter and only half-way towards its independent construction.

It is not until the fifth stage, i.e., when the object is sought in accordance with its successive displacements, that the group is really made objective, the combinativity of displacements, their reversibility and conservation of position ("identity") are achieved. Only the possibility of detours ("associativity") is still absent, for lack of sufficient anticipation, but it becomes more and more general during the sixth stage. Moreover, parallel with this progress, a system of relations between objects themselves is constructed, such as the relations "placed upon", "inside", or "outside", "in front of" or "behind" (with the correlating of distant planes with size constancy), etc.

We may thus conclude that the formation of the object's perceptual constancy by means of sensori-motor regulations goes hand in hand with the progressive construction of systems which are also sensori-motor but pass beyond the scope of perception and lead to a group structure—which, needless to say, is still exclusively practical and not conceptual. Why then does not perception itself also benefit from this structure and why does it remain at the level of simple regulations? The explanation for this is now clear; however "decentralised" it may be in relation to the initial centralisation of vision or of its particular organ, a perception is still egocentric and centred on an object in accordance with the subject's own perspective. Furthermore, the kind of decentralisation that characterises perception, i.e. co-ordination between successive centrings, arrives only at a composition of a statistical order, which is therefore incomplete (Chap. 3). Thus perceptual composition cannot rise above the level of what we have just been describing as the "subjective" group, i.e. a system centred with reference to the subject's own action, and capable, at the most, of corrections and regulations. And this is still true even at the stage at which the subject, passing beyond the perceptual field in order to anticipate and reconstruct invisible movements and objects, achieves an objectivised group structure in the realm of immediate "practical" space.

In general, we may thus conclude that there is an essential unity between the sensori-motor processes that engender perceptual activity, the formation of habits, and pre-verbal or pre-representative intelligence itself. The latter does not therefore arise as a new power, superimposed all of a sudden on completely prepared previous mechanisms, but is only the expression of these same mechanisms when they go beyond present and immediate contact with the world (perception), as well as beyond short and rapidly automatised connections between perceptions and responses (habit), and operate at progressively greater distances and by more complex routes, in the direction of mobility and reversibility. Early intelligence, therefore, is simply the form of mobile equilibrium towards which the mechanisms adapted to perception and habit tend; but the latter attain this only by leaving their respective fields of application. Moreover, intelligence, from this first sensori-motor stage onwards, has already succeeded in constructing, in the special case of space, the equilibrated structure that we call the group of displacements—in an entirely empirical or practical form, it is true, and of course remaining on the very restricted plane of immediate space. But it goes without saying that this organization, circumscribed as it is by the limitations of action, still does not constitute a form of thought. On the contrary, the whole development of thought, from the advent of language to the end of childhood, is necessary in order that the completed sensori-motor structures, which may even be co-ordinated in the form of empirical groups, may be extended into genuine operations, which will constitute or reconstruct these groupings and groups at the level of symbolic behaviour and reflective reasoning.

Part III

The Development of Thought

5

THE GROWTH OF THOUGHT—
INTUITION AND OPERATIONS

We have noted, in the first part of this work, that the operations of thought reach their form of equilibrium when they are formed into complex systems characterized by reversible combinativity (groupings or groups). But if a form of equilibrium marks the final limit of development, this does not explain either its initial phases or its constructive mechanism. In the second part, we were then able to locate the origin of operations in sensori-motor processes; the schemata of sensori-motor intelligence form the practical equivalent of concepts and relations, and their co-ordination into spatio-temporal systems of objects and movements even arrives, though still in a practical and empirical form, both at the conservation of the object, and at a correlative group structure (H. Poincaré's group of experienced displacements). But it is obvious that this sensori-motor group simply constitutes a schema of behaviour, i.e. the equilibrated system formed by the various possible physical movements in near space, and that it in no way attains the rank of an instrument of thought.[1] Certainly, sensori-motor intelligence lies at the source of thought, and continues to affect it throughout life through perceptions and practical sets. In particular, the role of perception in the most highly developed thought cannot be neglected, as it is by some writers when they pass too rapidly from neurology to sociology, and this role alone bears witness to the persistent influence of early schemata. But there is still a very long way to go from preverbal intelligence to operational thought before reflective groupings may be established, and even if there is a functional continuity between the two extremes, the formation of a series of intermediate structures at several heterogeneous levels is indispensable.

DIFFERENCES IN STRUCTURE BETWEEN CONCEPTUAL INTELLIGENCE AND SENSORI-MOTOR INTELLIGENCE

In order to understand the mechanism of the formation of operations, it is first of all important to realise what it is that has to be constructed, i.e. what must be added to sensori-motor intelligence for it to be extended into conceptual thought. Nothing indeed could be more superficial than to suppose that the construction of intelligence is already accomplished on the practical level, and then simply to appeal to language and imaginal representation to explain how this ready-made intelligence comes to be internalized as logical thought.

In point of fact, only the functional point of view allows us to find in sensori-motor intelligence the practical equivalent of

[1] If we divide behaviour into three main systems, organic hereditary structures (instinct), sensori-motor structures (which may be learned), and symbolic structures (which constitute thought), we may place the group of sensori-motor displacements at the apex of the second of these systems, while operational groups and groupings of a formal nature are at the top of the third.

classes, relations, reasonings and even groups of displacements in their empirical form as actual displacements. From the point of view of structure, and consequently of effect, there remain a certain number of fundamental differences between sensori-motor co-ordinations and conceptual co-ordinations, with regard both to the nature of the co-ordinations themselves and to the distances covered by the action, i.e. its scope of application.

In the first place, acts of sensori-motor intelligence, which consist solely in co-ordinating successive perceptions and (also successive) overt movements, can themselves only be reduced to a succession of states, linked by brief anticipations and reconstructions, but never arriving at an all-embracing representation; the latter can only be established if thought makes these states simultaneous, and thus releases them from the temporal sequence characteristic of action. In other words, sensori-motor intelligence acts like a slow-motion film, in which all the pictures are seen in succession but without fusion, and so without the continuous vision necessary for understanding the whole.

In the second place, and for the same reason, an act of sensori-motor intelligence leads only to practical satisfaction, i.e. to the success of the action, and not to knowledge as such. It does not aim at explanation or classification or taking note of facts for their own sake; it links causally and classifies and takes note of facts only in relation to a subjective goal which is foreign to the pursuit of truth. Sensori-motor intelligence is thus an intelligence in action and in no way reflective.

As regards its scope, sensori-motor intelligence deals only with real entities, and each of its actions thus involves only very short distances between subject and objects. It is doubtless capable of detours and reversals, but it never concerns anything but responses actually carried out and real objects. Thought alone breaks away from these short distances and physical pathways, so that it may seek to embrace the whole universe including what is invisible and sometimes even what cannot be pictured; this infinite expansion of spatio-temporal distances between subject and objects comprises the principal innovation of conceptual intelligence and the specific power that enables it to bring about operations.

There are thus three essential conditions for the transition from the sensori-motor level to the reflective level. Firstly, an increase in speed allowing the knowledge of the successive phases of an action to be moulded into one simultaneous whole. Next, an awareness, not simply of the desired results of action, but its actual mechanisms, thus enabling the search for the solution to be combined with a consciousness of its nature. Finally, an increase in distances, enabling actions affecting real entities to be extended by symbolic actions affecting symbolic representations and thus going beyond the limits of near space and time.

We see then that thought can neither be a translation nor even a simple continuation of sensori-motor processes in a symbolic form. It is much more than a matter of formulating or following up work already started; it is necessary from the start to reconstruct everything on a new plane. Perception and overt responses by themselves will continue to function in the same way, except for being charged with new meanings and integrated into new systems. But the structures of intelligence have to be entirely rebuilt before they can be completed; knowing how to reverse an object (cf. the bottle mentioned in Chap. 4) does not imply that one can represent a series of rotations in thought; physical movement along a complex route and returning to the starting-point does not necessarily involve understanding an imaginary system of displacements, and even to anticipate the conservation of an object in practice does not lead immediately to the conception of conservations affecting a system built up of different elements.

Moreover, in order to reconstruct these structures in thought, the subject is going to encounter the same difficulties, though transposed to this new level, that he has already overcome in immediate action. In order to construct a space, a time, a universe of causes and of sensori-motor or practical objects, the child has had to free himself from his perceptual and motor egocentricity; by a series of successive decentralisations he has managed to organise an empirical group of physical displacements, by localising his own body and his own movements amid the whole mass of others. This construction of groupings and operational groups of thought will necessitate a similar change of direction, but one following infinitely more complex paths. Thought will have to be decentralised, not only in relation to the perceptual centralisation of the movement, but also in relation to the whole of the subject's action. Thought, springing from action, is indeed egocentric at first for exactly the same reasons as sensori-motor intelligence is at first centred on the particular perceptions or movements from which it arises. The construction of transitive, associative and reversible operations will thus involve a conversion of this initial egocentricity into a system of relations and classes that are decentralised with respect to the self, and this intellectual decentralisation (not to mention its social aspect which we shall come back to in Chap. 6) will in fact occupy the whole of early childhood.

The development of thought will thus at first be marked by the repetition, in accordance with a vast system of loosenings and separations, of the development which seemed to have been completed at the sensori-motor level, before it spreads over a field which is infinitely wider in space and more flexible in time, to arrive finally at operational structures.

STAGES IN THE CONSTRUCTION OF OPERATIONS

In order to arrive at the mechanism of this development, which finds its final form of equilibrium in the operational grouping, we will distinguish (simplifying and schematizing the matter) four principal periods, following that characterized by the formation of sensori-motor intelligence.

After the appearance of language or, more precisely, the symbolic function that makes its acquisition possible ($1\frac{1}{2}$ − 2 years), there begins a period which lasts until nearly 4 years and sees the development of a symbolic and preconceptual thought.

From 4 to about 7 or 8 years, there is developed, as a closely linked continuation of the previous stage, an intuitive thought whose progressive articulations lead to the threshold of the operation.

From 7−8 to 11−12 years "concrete operations" are organized, i.e. operational groupings of thought concerning objects that can be manipulated or known through the senses.

Finally, from 11−12 years and during adolescence, formal thought is perfected and its groupings characterize the completion of reflective intelligence.

SYMBOLIC AND PRECONCEPTUAL THOUGHT

From the last stages of the sensori-motor period onwards, the child is capable of imitating certain words and attributing a vague meaning to them, but the systematic acquisition of language does not begin until about the end of the second year.

Now, direct observation of the child, as well as the analysis of certain speech disturbances, shows that the use of a system of verbal signs depends on the exercise of a more general "symbolic function", characterised by the representation of reality through the medium of "significants" which are distinct from "significates".

In fact, we should distinguish between symbols and signs on the one hand and indices or signals on the other. Not only all thought, but all cognitive and motor activity, from perception and habit to conceptual and reflective thought, consists in linking meanings, and all meaning implies a relation between a significant and a signified reality. But in the case of an index the significant constitutes a part or an objective aspect of the significate, or else it is linked to it by a causal relation; for the hunter tracks in the snow are an index of game, and for the infant the visible end of an almost completely hidden object is an index of its presence. Similarly, the signal, even when artificially produced by the experimenter, constitutes for the subject simply a partial aspect of the event that it heralds (in a conditioned response the signal is perceived as an objective antecedent). The symbol and the sign, on the other hand, imply a differentiation, from the point of view of the subject himself, between the significant and the significate; for a child playing at eating, a pebble representing a sweet is consciously recognized as that which symbolizes and the sweet as that which is symbolized; and when the same child, by "adherence to the sign", regards a name as inherent in the thing named, he nevertheless regards this name as a significant, as though he sees it as a label attached in substance to the designated object.

We may further specify that, according to a custom in linguistics which may usefully be employed in psychology, a symbol is defined as implying a bond of similarity between the significant and the significate, while the sign is "arbitrary" and of necessity based on convention. The sign thus cannot exist without social life, while the symbol may be formed by the individual in isolation (as in young children's play). Of course symbols also may be socialized, a collective symbol being generally half sign and half symbol; on the other hand, a pure sign is always collective.[1]

[1] This proposed terminology may conflict with existing usage in English. For example, C. R. Morris (in *Signs, Language and Behavior*, New York: Prentice Hall 1946), uses *symbol* to mean any *sign* produced by an interpreter and acting as a substitute for another *sign* with which it is synonymous. All *signs* which are not *symbols* are *signals*. Morris's *iconic signs* and *lansigns* (or *language-signs*) appear to approximate to Piaget's *symbols* and *signs* respectively. (*Translator's note.*)

In view of this, it should be noted that the acquisition of language, i.e. the system of collective signs, in the child coincides with the formation of the symbol, i.e. the system of individual significants. In fact, we cannot properly speak of symbolic play during the sensori-motor period, and K. Groos has gone rather too far in attributing an awareness of make-believe to animals. Primitive play is simply a form of exercise and the true symbol appears only when an object or a gesture represents to the subject himself something other than perceptible data. Accordingly we note the appearance, at the sixth of the stages of sensori-motor intelligence, of "symbolic schemata," i.e. schemata of action removed from their context and evoking an absent situation (e.g. pretending to sleep). But the symbol itself appears only when we have representation dissociated from the subject's own action: e.g. putting a doll or a teddy-bear to bed. Now precisely at the stage at which the symbol in the strict sense appears in play, speech brings about in addition the understanding of signs.

As for the formation of the individual symbol, this is elucidated by the development of imitation. During the sensori-motor period, imitation is only an extension of the accommodation characteristic of assimilatory schemata. When he can execute a movement, the subject, on perceiving an analogous movement (in other persons or in objects), assimilates it to his own, and this assimilation, being as much motor as perceptual, activates the appropriate schema. Subsequently, the new instance elicits an analogous assimilatory response, but the schema activated is then accommodated to new details; at the sixth stage, this imitative assimilation becomes possible even with a delay, thus presaging representation. Truly representative imitation, on the other hand, only begins with symbolic play because, like the latter, it presupposes imagery. But is the image the cause or the effect of this internalization of the imitative mechanism? The mental image is not a primary fact, as associationism long believed; like

imitation itself, it is an accommodation of sensori-motor schemata, i.e. an active copy and not a trace or a sensory residue of
perceived objects. It is thus internal imitation and is an extension
of the accommodatory function of the schemata characteristic of
perceptual activity (as opposed to perception itself), just as the
external imitation found at previous levels is an extension of the
accommodatory function of sensori-motor schemata (which are
closely bound up with perceptual activity).

From then on, the formation of the symbol may be explained
as follows: deferred imitation, i.e. accommodation extended in
the form of imitative sketches, provides significants, which play
or intelligence applies to various significates in accordance with
the free or adapted modes of assimilation that characterize these
responses. Symbolic play thus always involves an element of imitation functioning as a significant, and early intelligence utilises
the image in like manner, as a symbol or significant.[1]

We can understand now why speech (which is likewise
learned by imitation, but by an imitation of ready-made signs,
whereas imitation of shapes, etc., provides the significant
material of private symbolism) is acquired at the same time as
the symbol is established: it is because the use of signs, like that
of symbols, involves an ability which is quite new with respect
to sensori-motor behaviour and consists in representing one
thing by another. We may thus apply to the infant this idea of a
general "symbolic function", which has sometimes been used as
a hypothesis in connection with aphasia, since the formation of
such a mechanism is believed, in short, to characterize the simultaneous appearance of representative imitation, symbolic play,
imaginal representation and verbal thought.

To sum up, the beginnings of thought, while carrying on the
work of sensori-motor intelligence, spring from a capacity for
distinguishing significants and significates, and consequently

---

[1] See I. Meyerson, "Les images" in Dumas, *Nouveau Traité de Psychologie*.

rely both on the invention of symbols and on the discovery of signs. But needless to say, for a young child who finds the system of ready-made collective signs inadequate, since they are partly inaccessible and are hard to master, these verbal signs will for a long time remain unsuitable for the expression of the particular entities on which the subject is still concentrated. This is why, as long as egocentric assimilation of reality to the subject's own action prevails, the child will require symbols; hence symbolic play or imaginative play, the purest form of egocentric and symbolic thought, the assimilation of reality to the subject's own interests and the expression of reality through the use of images fashioned by himself.

But even in the field of applied thought, i.e. the beginnings of representative intelligence, tied more or less closely to verbal signs, it is important to note the role of imaginal symbols and to realize how far the subject is, during his early childhood, from arriving at genuine concepts. We must, in fact, distinguish a first period in the development of thought, lasting from the appearance of language to the age of about 4 years, which may be called the period of preconceptual intelligence and which is characterized by preconcepts or participations and, in the first forms of reasoning, by "transduction" or preconceptual reasoning.

Pre-concepts are the notions which the child attaches to the first verbal signs he learns to use. The distinguishing characteristic of these schemata is that they remain midway between the generality of the concept and the individuality of the elements composing it, without arriving either at the one or at the other. The child aged 2–3 years will be just as likely to say "slug" as "slugs" and "the moon" as "the moons", without deciding whether the slugs encountered in the course of a single walk or the discs seen at different times in the sky are one individual, a single slug or moon, or a class of distinct individuals. On the one hand, he cannot yet cope with general classes, being unable to distinguish between "all" and "some". On the other hand, although the idea of the permanent individual object has been formed in the field of immediate action, such is by no means the case where distant space and reappearances at intervals are concerned; a mountain is still deemed to change its shape in the course of a journey (just as in the earlier case of the rotated feeding-bottle) and "the slug" reappears in different places. Hence, sometimes we have true "participations" between objects which are distinct and distant from each other: even at the age of four years, a shadow, thrown on a table in a closed room by means of a screen, is explained in terms of those which are found "under the trees in the garden" or at night-time, etc., as though these intervened directly the moment the screen is placed on the table (and with the subject making no attempt to go into the "how" of the phenomenon).

It is clear that such a schema, remaining midway between the individual and the general, is not yet a logical concept and is still partly something of a pattern of action and of sensori-motor assimilation. But it is nevertheless a representative schema and one which, in particular, succeeds in evoking a large number of objects by means of privileged elements, regarded as samples of the pre-conceptual collection. On the other hand, since these type-individuals are themselves made concrete by images as much as, and more than, by words, the pre-concept improves on the symbol in so far as it appeals to generic samples of this kind. To sum up then, it is a schema placed midway between the sensori-motor schema and the concept with respect to its manner of assimilation, and partaking of the nature of the imaginal symbol as far as its representative structure is concerned.

Now the reasoning that consists in linking such preconcepts shows precisely the same structures. Stern gave the name "transduction" to these primitive reasonings, which are effected not by deduction but by direct analogies. But that is not quite all:

pre-conceptual reasoning or transduction is based only on incomplete dovetailings and is thus inadequate for any reversible operational structure. Moreover, if it succeeds in practice, it is because it merely consists of a sequence of actions symbolized in thought, a true "mental experiment", i.e. an internal imitation of actions and their results, with all the limitations that this kind of empiricism of the imagination involves. We thus see in transduction both the lack of generality that is inherent in the preconcept and its symbolic or imaginal character which enables actions to be transposed into thought.

INTUITIVE THOUGHT

The forms of thought we have been describing can be analysed only through observation, since young children's intelligence is still far too unstable for them to be interrogated profitably. After about 4 years, on the other hand, short experiments with the subject, in which he has to manipulate experimental objects, enable us to obtain regular answers and to converse with him. This fact alone indicates a new structuring.

In fact, from 4 to 7 years we see a gradual co-ordination of representative relations and thus a growing conceptualization, which leads the child from the symbolic or preconceptual phase to the beginnings of the operation. But the remarkable thing is that this intelligence, whose progress may be observed and is often rapid, still remains pre-logical even when it attains its maximum degree of adaptation;[1] up to the time when this series of successive equilibrations culminates in the "grouping", it continues to supplement incomplete operations with a semi-symbolic form of thought, i.e. intuitive reasoning; and it controls judgments solely by means of intuitive "regulations",

---

[1] We are disregarding here purely verbal forms of thought, such as animism, infantile artificialism, nominal realism, etc.

which are analogous on a representative level to perceptual adjustments on the sensori-motor plane.

As an example let us consider an experiment which we conducted some time ago with A. Szeminska. Two small glasses, A and A_2, of identical shape and size, are each filled with an equal number of beads, and this equality is acknowledged by the child, who has filled the glasses himself, e.g. by placing a bead in A with one hand every time he places a bead in A_2 with the other hand. Next, A_2 is emptied into a differently shaped glass B, while A is left as a standard. Children of 4–5 years then conclude that the quantity of beads has changed, even though they are sure none has been removed or added. If the glass B is tall and thin they will say that there are "more beads than before" because "it is higher", or that there are fewer because "it is thinner", but they agree on the non-conservation of the whole.

First, let us note the continuity of this reaction with those of earlier levels. The subject possesses the notion of an individual object's conservation but does not yet credit a collection of objects with permanence. Thus, the unified class has not been constructed, since it is not always constant, and this non-conservation is an extension both of the subject's initial reactions to the object (with a greater flexibility due to the fact that it is no longer a question of an isolated element but of a collection) and of the absence of an understanding of plurality which we mentioned in connection with the pre-concept. Moreover, it is clear that the reasons for the error are of a quasi-perceptual order; the rise in the level, or the thinness of the column, etc., deceives the child. However, it is not a question of perceptual illusions; perception of relations is on the whole correct, but it occasions an incomplete intellectual construction. It is this pre-logical schematization, which is still closely modelled on perceptual data though it recentres them in its own fashion, that may be called intuitive thought. We can see straight away how it is related to the imaginal character of the pre-concept and to the mental experiments that characterize transductive reasoning.

However, this intuitive thought is an advance on preconceptual or symbolic thought. Intuition, being concerned essentially with complex configurations and no longer with simple half-individual, half-generic figures, leads to a rudimentary logic, but in the form of representative regulations and not yet of operations. From this point of view, there exist intuitive "centralisations" and "decentralisations" which are analogous to the mechanisms we mentioned in connection with the sensori-motor schemata of perception (Chap. 3). Suppose a child estimates that there are more beads in B than in A because the level has been raised. He thus "centres" his thought, or his attention,[1] on the relation between the heights of B and A, and ignores the widths. But let us empty B into glasses C or D, etc., which are even thinner and taller; there must come a point at which the child will reply, "there are fewer, because it is too narrow". There will thus be a correction of centring on height by a decentring of attention on to width. On the other hand, in the case of the subject who estimates the quantity in B as less than that in A on account of thinness, the lengthening of the column in C, D, etc., will induce him to reverse his judgment in favour of height. Now this transition from a single centring to two successive centrings heralds the beginnings of the operation; once he reasons with respect to both relations at the same time, the child will, in fact, deduce conservation. However, in the case we are considering, there is neither deduction nor a true operation; an error is simply corrected, but it is corrected late and as a reaction to its very exaggeration (as in the field of perceptual illusions), and the two relations are seen alternately instead of being logically multiplied. So all that occurs is a kind of intuitive regulation and not a truly operational mechanism.

[1] Concentration of attention on one idea is precisely nothing else but the centring of thought.

That is not all. In studying the differences between intuition and operation together with the transition from the one to the other, we may consider not merely the relating to each other of qualities forming two dimensions but their correspondences in either a logical (i.e. qualitative) or a mathematical form. The subject is first presented with glasses A and B of different shapes and he is asked to place a bead simultaneously in each glass, one with the left hand and one with the right. With small numbers (4 or 5), the child immediately believes in the equivalence of the two collections, which seems to presage the operation, but when the shapes change too much, even though the one-to-one correspondence is continued, he ceases to recognize equality. The latent operation is thus destroyed by the deceptive demands of intuition.

Let us line up six red counters on a table, supply the subject with a collection of blue counters and ask him to place on the table as many blue ones as there are red ones. From about 4 to 5 years, the child does not establish any correspondence and contents himself with a row of equal length (with its members closer together than those of the standard). At about 5 or 6 years, on the average, the subject lines up six counters opposite the blue. Has the operation now been acquired, as might appear? Not at all! It is only necessary to spread the elements in one of the series further apart, or to draw them close together, etc. for the subject to disbelieve in the equivalance. As long as the optical correspondence lasts, the equivalence is obvious; once the first is changed, the second disappears, which brings us back to the non-conservation of the whole.

Now this intermediate reaction is full of interest. The intuitive schema has become flexible enough to enable a correct system of correspondences to be anticipated and constructed, which to an uninformed observer presents all the appearances of an operation. And yet, once the intuitive schema is modified, the logical relation of equivalence, which would be the necessary product of an operation, is shown not to have existed. We are thus confronted with a form of intuition which is superior to that of the previous level and which may be called "articulated intuition" as opposed to simple intuition. But this articulated intuition, although it approaches the operation (and eventually joins up with it by stages which are often imperceptible), is still rigid and irreversible like all intuitive thought; it is thus only the product of successive regulations which have finally articulated the original global and unanalysable relations, and it is not yet a genuine "grouping".

This difference between the intuitive and the operational methods may be pinned down still further by directing the analysis towards the formation of classes and the seriation of asymmetrical relations, which constitute the most elementary groupings. But of course the problem must be presented on an intuitive plane, the only one accessible at this stage, as opposed to a formal plane indissociably tied to language. To study the formation of classes, we place about twenty beads in a box, the subject acknowledging that they are "all made of wood", so that they constitute a whole, B. Most of these beads are brown and constitute part A, and some are white, forming the complementary part A'. In order to determine whether the child is capable of understanding the operation $A + A' = B$, i.e. the uniting of parts in a whole, we may put the following simple question: In this box (all the beads still being visible) which are there more of—wooden beads or brown beads, i.e. is $A < B$?

Now, up to about the age of 7 years, the child almost always replies that there are more brown beads "because there are only two or three white ones". We then question further: "Are all the brown ones made of wood?"—"Yes."—"If I take away all the wooden beads and put them here (a second box) will there be any beads left in the (first) box?"—"No, because they are all made of wood."—"If I take away the brown ones, will there be any beads left?"—'Yes, the white ones." Then the original question is repeated and the subject continues to state that there are more brown beads than wooden ones in the box because there are only two white ones, etc.

The mechanism of this type of reaction is easy to unravel: the subject finds no difficulty in concentrating his attention on the whole B, or on the parts A and A', if they have been isolated in thought, but the difficulty is that by centring on A he destroys the whole, B, so that the part A can no longer be compared with the other part A'. So there is again a non-conservation of the whole for lack of mobility in the successive centralisations of thought. But this is still not all. When the child is asked to imagine what would happen if we made a necklace either with the wooden beads or with the brown beads, A, we again meet the foregoing difficulties but with the following details: "If I make a necklace with the brown ones", a child will sometimes reply, "I could not make another necklace with the same beads, and the necklace made of wooden beads would have only white ones!" This type of thinking, which is in no way irrational, nevertheless shows the difference still separating intuitive thought and operational thought. In so far as the first imitates true actions by imagined mental experiments, it meets with a particular obstacle, namely, that in practice one could not construct two necklaces at the same time from the same elements, whereas in so far as the second is carried out through internalized actions that have become completely reversible, there is nothing to prevent two hypotheses being made simultaneously and then being compared with each other.

The seriation of sticks A, B, C, etc. of different lengths, but placed side by side (to be compared in pairs), also yields an interesting lesson. Children of 4 to 5 years are able to construct only unco-ordinated pairs, BD, AC, EG, etc. Then the child constructs short series and achieves the seriation of ten elements only by groping his way from step to step. Furthermore, when he has finished a row he is incapable of interpolating new terms without undoing the whole. Not until the operational level is seriation achieved straight away, by such a method as, for example, finding the smallest of all the terms and then the next smallest, etc. It is at this level, similarly, that the inference $(A < B) + (B < C) (A < C)$ becomes possible, whereas at intuitive levels the subject declines to derive from the two perceptually verified inequalities $A < B$ and $B < C$ the conclusion $A < C$.

The progressive articulations of intuition and the differences which still separate them from the operation are particularly clear where space and time are concerned, as well as being very instructive owing to the possibility of comparing intuitive and sensori-motor reactions. We are thus reminded of how the infant learns the action of turning a bottle round. To reverse an object by an intelligent action does not automatically lead to knowing how to reverse it in thought, and the stages of this intuition of rotation constitute largely a repetition of those of actual or sensori-motor rotation; in both cases we find a similar process of progressive decentralisation from the egocentric point of view, this decentralisation being simply perceptual and motor in the first case, and representative in the second.

In this connection we may proceed in two ways, either by moving the subject (in thought) around the object, or else by rotating the object itself in thought. To bring about the first situation, we may, for example, show the child cardboard mountains on a square table and ask him to choose from several very simple drawings those which correspond to possible points of view (the child, sitting at one side of the table, sees a doll move round the mountain and has to pick out the pictures that correspond to its different positions); now small children still remain dominated by the point of view that is theirs at the moment of choice, even when they have previously walked round the table from one side to the other. Reversals from front to behind and from left to right are difficulties which are at first insurmountable and only acquired gradually by intuitive regulations up to 7 or 8 years.

The rotation of the object itself, on the other hand, provides interesting data concerning the intuition of order. For example, three mannikins of different colours, A, B and C, are threaded on a single wire, or else three balls, A, B and C, are placed into a cardboard tube (so constructed that the balls cannot change their relative positions). The child is required to draw the whole as an aid to memory. Then the elements A, B and C are moved behind a screen or through the tube and the child has to predict the order in which they will emerge at the other end (i.e. their original order) and the opposite order of emergence when they return. All children foresee the original order. The opposite order, on the other hand, is beyond them till about 4 or 5 years, the end of the pre-conceptual period. Next, the whole apparatus (tube or wire) is turned through 180 degrees and the subject has to predict the order of emergence (which is thus reversed). After the child has himself checked the result we begin again and execute two half-circles (360 degrees in all), then three, etc.

This demonstration enables us to follow the whole progress of intuition step by step right up to the beginnings of the operation. From 4 to 7 years of age the subject is unable to foresee that half a turn will change the order A B C into C B A; then, having put the matter to the test, he admits that two half-turns will actually produce C B A. Although undeceived by experience, he is no better able to predict the effect of three half-turns. Moreover, children of 4 to 5 years, after seeing sometimes A and sometimes C at the head of the column, imagine that B also will have its turn as leader (not knowing Hilbert's axiom, according to which, if B is "between" A and C, it must also necessarily be "between" C and A!) The idea of the invariability of the "between" position is also acquired by the successive regulations that are responsible for the articulation of intuition. Not until about 7 is the whole system of changes understood, and often this last phase is rather sudden on account of a general "grouping" of the relations involved. It should be noted straight away that the operation thus follows from intuition not merely when the original order (+) can be reversed in thought (−) by a primary intuitive articulation but even when two opposite orders yield the original order again (− multiplied by − gives +, which in this particular case is understood at 7−8 years!)

Temporal relations provide similar data. Intuitive time is a time which is tied to particular objects or movements and which has no homogeneity or uniform flow. When two moving objects leave the same point A and arrive at two different places, B and B′, the 4−5 year-old child acknowledges the simultaneity of the departures but usually contests that of the arrivals, although this is easily perceptible. He recognises that one of the objects ceased to move when the other stopped, but he refuses to grant that the movements ceased "at the same time", because there simply is as yet no time common to different speeds. Similarly, he conceives of "before" and "after" in terms of spatial succession and not yet in terms of temporal succession. From the point of view of duration, "faster" implies "more time" even in the absence of verbal implication, and simply by inspection of the data (since faster = further = more time). When these first difficulties have been overcome by an articulation of intuitions (due to decentralisation of thought, which becomes accustomed to comparing two systems of positions at the same time, whence a gradual regulation of estimations), there nevertheless still exists a systematic incapacity to combine local times into one single time. Two equal quantities of water flowing, at the same rate through the two branches of a tube into differently shaped bottles, give rise, for example, to the following judgments: the 6−7 years old child recognizes the simultaneity of starts and stops but denies that water has been flowing into one bottle for as long as it has flowed into the other. Ideas concerning age give rise to similar statements; if A was born before B, that does not mean that he is older and, if he is older, that does not exclude the possibility that B might catch up with him or even overtake him!

These intuitive ideas are parallel to those encountered in the field of practical intelligence. André Rey has shown how subjects of the same age, tackling problems involving instrumental devices (extracting objects from a tube with hooks, changing round plugs, rotations, etc), also show irrational behaviour before these adaptive solutions are discovered.[1] With regard to representations without manipulation, such as the explanation of the movement of rivers or clouds, the floating of boats, etc., we have shown that causal links of this type were based on bodily action; physical movement implies teleology, an active internal force; the river "leaps" over pebbles, the clouds make the wind, which in turn pushes them, and so on.[2]

This then is intuitive thought. Like symbolic thought of a preconceptual nature, from which it springs directly, it is, in a sense, an extension of sensori-motor intelligence. Just as the latter assimilates objects to response-schemata, so intuition is always in the first place a kind of action carried out in thought; pouring from one vessel to another, establishing a correspondence, joining, serialising, displacing, etc. are still response-schemata to which representation assimilates reality. But the accommodation of these schemata to objects, instead of remaining practical, provides imitation or imaginal significants which enable this same assimilation to occur in thought. So in the second place, intuition is an imaginal thought, more refined than that of the previous period, since it concerns complex configurations and not merely simple syncretic collections symbolized by type-individuals; but it still uses representative

[1] André Rey, *L'Intelligence pratique chez l'enfant*, Alcan, 1935.
[2] *La Causalité physique chez l'enfant*, Alcan, 1927.

symbolism and therefore constantly exhibits some of the limitations that are inherent in this.

These limitations are obvious. Intuition, being a direct relationship between a schema of internalized action and the perception of objects, results only in configurations "centred" on this relationship. Since it is unable to go beyond these imaginal configurations, the relations that it constructs are thus incapable of being combined. The subject does not arrive at reversibility, because an action translated into a simple imagined experiment is still uni-directional, and because an assimilation centred on a perceptual configuration is necessarily uni-directional also. Hence the absence of transitivity, since each centring distorts or destroys the others, and of associativity, since the relations vary with the route followed by thought in fashioning them. Altogether then, in the absence of transitive, reversible and associative combinativity, there is neither a guarantee of the identity of elements nor a conservation of the whole. Thus, we may also say that intuition is still phenomenalist, because it copies the outlines of reality without correcting them, and egocentric, because it is constantly related to present action; in this way, it lacks an equilibrium between the assimilation of phenomena to thought-schemata and the accommodation of the latter to reality.

But this initial state, which recurs in each of the fields of intuitive thought, is progressively corrected, thanks to a system of regulations which herald operations. Intuition, at first dominated by the immediate relations between the phenomenon and the subject's viewpoint, evolves towards decentralisation. Each distortion, when carried to an extreme, involves the re-emergence of the relations previously ignored. Each relation established favours the possibility of a reversal. Each detour leads to interactions which supplement the various points of view. Every decentralisation of an intuition thus takes the form of a regulation, which is a move towards reversibility, transitive combinativity and associativity, and thus, in short, to conservation through the co-ordination of different viewpoints. Hence we have articulated intuitions, which progress towards reversible mobility and pave the way for the operation.

CONCRETE OPERATIONS

The appearance of logico-arithmetical and spatio-temporal operations introduces a problem of considerable interest in connection with the mechanisms characterising the development of thought. The point at which articulated intuitions turn into operational systems is not to be determined by mere convention, based on definitions decided on in advance. To divide developmental continuity into stages recognizable by some set of external criteria is not the most profitable of occupations; the crucial turning-point for the beginning of operations shows itself in a kind of equilibration, which is always rapid and sometimes sudden, which affects the complex of ideas forming a single system and which needs explaining on its own account. In this there is something comparable to the abrupt complex restructurings described in the Gestalt theory, except that, when it occurs, there arises the very opposite of a crystallisation embracing all relations in a single static network; operations, on the contrary, are found formed by a kind of thawing out of intuitive structures, by the sudden mobility which animates and co-ordinates the configurations that were hitherto more or less rigid despite their progressive articulation. Thus, quite distinct stages in development are marked, for example, by the point at which temporal relations are merged in the notion of a single time, or the point at which the elements of a complex are conceived as constituting an unvarying whole or the inequalities characterising a system of relations are serialised in a single scale, and so on; after trial-and-error imagination there follows, sometimes abruptly, a feeling of coherence and of necessity, the satisfaction of arriving at a system which is both complete in itself and indefinitely extensible.

Consequently, the problem is to understand what internal process effects this transition from a phase of progressive equilibration (intuitive thought) to a mobile equilibrium which is reached, as it were, at the limit of the former (operations). If the concept of "grouping" described in Chapter 2 has, in fact, a psychological meaning, this is precisely the point at which it should reveal it.

So, assuming that the intuitive relations of a given system are at a certain moment suddenly "grouped", the first question is to decide by what internal or mental criterion grouping is to be recognised. The answer is obvious: where there is "grouping" there will be the conservation of a whole, and this conservation itself will not merely be assumed by the subject by virtue of a probable induction, but affirmed by him as a certainty in his thought.

In this connection let us reconsider the first example cited with reference to intuitive thought: the pouring of the beads from one glass to another. After a long period during which each pouring out is believed to change the quantities, and after an intermediate phase (articulated intuition) when some transfers are believed to change the whole while others, between glasses that are just slightly different, induce the subject to suppose that the whole is conserved, there always comes a time (between 6 years and 7 years 8 months) when the child's attitude changes: he no longer needs to reflect, he decides, he even looks surprised that the question is asked, he is *certain* of the conservation. What has happened then? If we ask him his reasons, he replies that nothing has been removed or added; but the younger children also are well aware of this, and yet they do not infer identity. Thus, in spite of what E. Meyerson says, identification is not a primary process but the result of an assimilation by the whole grouping (the product of the original operation multiplied by its converse). Or else he replies that the height makes up for the width lost by the new glass, etc., but articulated intuition has already led to these decentrings of a given relation without their resulting in the simultaneous co-ordination of relations or in their necessary conservation. Or else, and this especially, he replies that a transfer from A to B may be corrected by a transfer from B to A and this reversibility is certainly essential, but the younger children have already on occasion admitted the possibility of a return to the starting-point, without this "empirical reversal" yet constituting a complete reversibility. There is, therefore, only one legitimate answer: the various transformations involved—reversibility, combination of compensated relations, identity, etc.—in fact depend on each other and, because they amalgamate into an organised whole, each is really new despite its affinity with the corresponding intuitive relation that was already formed at the previous level.

Let us take another example. In the case of the elements arranged in the order ABC and subjected to a half-rotation (180 degrees), the child intuitively and gradually discovers almost all the relationships: that B invariably remains "between" A and C and between C and A; that one half-turn changes ABC into CBA and that two half-turns lead back to ABC, etc. But the relation-ships discovered one after another are still intuitions with no link between them or "necessity" about them. At about 7 or 8 years, on the other hand, we find subjects who, before any trial, foresee:

1. that ABC reversed is CBA;
2. that two reversals result in the original order;
3. that three reversals are equivalent to one, etc.

Here again, each of the relationships may correspond to an intuitive discovery, but together they constitute a new reality, since they have become deductive and no longer consist of a succession of actual or mental experiments.

Now it is easy to see that in all such cases—and they are innumerable—a mobile equilibrium is reached when the following changes are simultaneously effected:

1. two successive actions can be combined into one;
2. the action-schema already at work in intuitive thought becomes reversible;
3. the same point can be reached by two different paths without being altered;
4. a return to the starting-point finds the starting-point unchanged;
5. when the same action is repeated, it either adds nothing to itself or else is a new action with a cumulative effect. In these we recognize transitive combinativity, reversibility, associativity and identity, with (in 5) either logical tautology or numerical iteration, all of which characterize logical "groupings" or arithmetical "groups".

But what must be clearly understood if we are to arrive at the true psychological nature of the grouping, as distinct from its formulation in logical language, is that these various closely related changes are actually the expression of one and the same total act, namely, an act of complete decentralisation or complete conversion of thought. The distinguishing characteristic of the sensori-motor schema (perception, etc.), of the preconceptual symbol and also of the intuitive configuration, is that they are always "centred" on a particular state of the object and a point of view peculiar to the subject; thus they always testify both to an egocentric assimilation to the subject and to a phenomenalist accommodation to the object. On the other hand, the distinguishing characteristic of the mobile equilibrium peculiar to the grouping is that the decentralisation, already provided for by the progressive regulations and articulations of intuition, suddenly becomes systematic on reaching its limit; thought is then no longer tied to particular states of the object, but is obliged to follow successive changes with all their possible detours and reversals; and it no longer issues from a particular viewpoint of the subject, but co-ordinates all the different viewpoints in a system of objective reciprocities. The grouping thus realizes for the first time an equilibrium between the assimilation of objects to the subject's action and the accommodation of subjective schemata to modifications of objects. At the outset, in fact, assimilation and accommodation act in opposite directions; hence the distorting character of the first and the phenomenalist character of the second. By means of anticipations and reconstitutions, which extend action in both directions to ever increasing distances, from the brief anticipations and reconstitutions characteristic of perception, habit and sensori-motor intelligence to the anticipatory schemata formed by intuitive representation, assimilation and accommodation are gradually equilibrated. The completion of this equilibrium explains the reversibility which is the final term of sensori-motor and mental anticipations and reconstitutions, and with it the reversible combinativity which is the distinguishing mark of the grouping; the detailed working of operations simply expresses, in fact, the combined conditions of a co-ordination of successive viewpoints of the subject (with possible reversal in time and anticipation of their sequel) and a co-ordination of perceptible or representable modifications of objects (in the past, in the present, or in the course of subsequent events).

In practice, the operational groupings, which are constituted at about 7 or 8 years of age (sometimes a little earlier), lead to the following structures. First of all they lead to the logical operations of fitting classes together (the problem of the brown beads, A, being less numerous than the wooden beads, B, is solved at about the age of 7) and of serialising asymmetrical relations. Hence the discovery of transitivity which permits of the deductions: $A = B$, $B = C$, therefore $A = C$; or $A < B$, $B < C$ therefore $A < C$. Furthermore, as soon as these additive groupings are acquired, multiplicative groupings are immediately understood in the form of correspondences: if he knows how to serialise objects according to the relations $A1 < B1 < C1$ ... the subject will find it no more difficult to serialise two or more sets, such as $A2 < B2 < C2$ ..., which correspond to each other term for term; when a child aged 7 has arranged a series of mannikins in order of size, he will be able to make a series of sticks or bags correspond to them, and he will be able to identify which element in one series corresponds to which in another even when they are all jumbled (since the multiplicative character of this grouping adds no difficulty to the additive serialising operations which have just been discovered).

Moreover, the simultaneous construction of the groupings of classification and of qualitative seriation means the advent of the system of numbers. Doubtless the young child does not have to wait for this operational generalisation to construct the first numbers (according to A. Descoedres, a new number is learnt each year from 1 to 5 years), but the numbers 1 to 6 are still intuitive since they are bound to perceptual configurations. Similarly, the child may be taught to count, but experiment reveals that the verbal use of the names of numbers has little connection with numerical operations as such, which sometimes precede counting aloud and sometimes follow it, with no necessary bond between the two. All that the operations constituting number, i.e. one-to-one correspondence (with conservation of the resulting equivalence despite differences in shape), or simple repetition of unity ($1 + 1 = 2$; $2 + 1 = 3$; etc.) require are the additive groupings of classification and of the serialisation of asymmetrical relations (ordering); but these are blended into a single operational whole, so that the unit 1 is simultaneously an element in a class (1 included in 2, 2 in 3, etc.), and in a series (the first 1 preceding the second 1, etc.). As long as the subject sees the individual elements with all their qualitative diversity, he can in fact either combine them according to their equivalent qualities (he then constructs classes) or arrange them according to their differences (he then constructs asymmetrical relations), but he cannot group them simultaneously as equivalent and different. Number, on the other hand, is a collection of objects conceived as both equivalent and orderable, their only difference thus being reduced to their position in a series. This combination of difference and equivalence implies, in this case, the elimination of quality, and that is precisely what accounts for the formation of the homogeneous unit 1 and the transition from logic to mathematics. Now it is very interesting to observe that this transition occurs just when logical operations are being constructed; classes, relations and numbers thus form a psychologically and logically indivisible whole, in which each of the three terms completes the other two.

But these logico-arithmetical operations constitute only one aspect of the fundamental groupings whose construction characterises the age, on the average, of 7–8 years. Corresponding to these operations, which assemble objects in order to classify, serialise or number them, are the operations that constitute objects themselves, complex yet unique objects such as space, time and material systems. Now, it is not surprising that these infra-logical or spatio-temporal operations are grouped in correlation with logico-arithmetical operations, since they are the same operations but on another scale: the joining together of objects in classes and of classes with one another becomes the joining of parts or pieces in a whole; seriation expressing differences between objects appears as relations of order (placing operations) and displacement, and number corresponds to measurement. Now it so happens that while classes, relations and numbers are being formed, we can see the construction, in a remarkably parallel manner, of the qualitative groupings that generate time and space. At the age of about 8, the relations of temporal order (before and after) are co-ordinated with duration (longer or shorter length of time), whereas the two systems of ideas were still independent at the intuitive level; as soon as they become joined in a single whole they engender the notion of a time common to various movements (internal and external) at different velocities. Above all, there are also constituted at the age of about 7 or 8 the qualitative operations that structure space: the spatial order of succession and the joining together of intervals or distances; conservation of lengths, areas, etc.; formation of a system of co-ordinates; perspectives and sections, etc. In this connection, the study of the spontaneous measurement that derives from early estimation by perceptual "transportation" and leads, at the age of 7 or 8, to the transitivity of operational equivalences (A = B, B = C, therefore A = C) and to the formation of the unit (by a synthesis of division and displacement), demonstrates in the clearest possible way how the continuous development first of perceptual and then of intuitive acquisitions leads finally to reversible operations as their necessary form of equilibrium.

But it is important to note that these different logico-arithmetical or spatio-temporal groupings are as yet far from constituting a formal logic applicable to all ideas and to all reasoning. This is an essential point which must be stressed, for the sake both of the theory of intelligence and of its educational applications, if we wish to adapt teaching to the findings of developmental psychology as opposed to the logical bias of scholastic tradition. In fact, the same children as reach the operations just described are usually incapable of them when they cease to manipulate objects and are invited to reason with simple verbal propositions. The operations that are involved here, then, are "concrete operations" and not yet formal ones; being constantly tied to action, they give it a logical structure, embracing also the speech accompanying it, but they by no means imply the possibility of constructing a logical discourse independently of action. Thus, class-inclusion is understood in the concrete problem of the beads (see above) from the age of 7–8 years, while a verbal test of identical structure is not solved until much later (cf. one of Burt's tests : "Some of the flowers in my bunch are yellow," says a boy to his sisters. The first replies, "Then all your flowers are yellow," the second replies, "Some of them are yellow", and the third: "None". Who is right?)

But this is not yet all. The same "concrete" inferences, such as those leading to the conservation of the whole, to transitivity of equality $(A = B = C)$ or of differences $(A < B < C \ldots)$, may be easily handled in the case of one particular system of ideas (such as quantity of material) and yet be meaningless for the same subjects in the case of another system of ideas (such as weight). In view of this especially, it is wrong to speak of formal logic before the end of childhood. "Groupings" are still relative to the types of concrete ideas (i.e. internalised actions) that they have actually structured, but the structuring of other types of concrete ideas, which are of a more complex intuitive nature, since they depend on quite different actions, requires a reconstruction of the same groupings independently of time.

A particularly clear example is the notion of the conservation of the whole (which is the very hall-mark of the grouping). Thus, the subject is given two pellets of dough to be modelled into the same shape, size and weight, then one of them is modified (made into a roll, etc.) and the subject is asked whether the material (the quantity of dough), the weight and the volume remain the same (volume is estimated by the displacement of water in two glasses in which the objects are immersed). Now after the age of 7 or 8 the quantity of material is recognised as conserved of necessity by virtue of the inferences already described in connection with the conservation of complexes. But up to 9–10 years the same subjects dispute that weight is conserved, and this comes of relying on the intuitive inferences that they used before 7–8 years to ascribe non-conservation to the material. As for the inferences they have just used (often only a few moments earlier) to prove the conservation of substance, this is not applied to weight at all. If the roll is thinner than the pellet the material is conserved, because this narrowing is compensated by lengthening, but the weight is reduced because, so it is held, narrowing acts unconditionally! At about 9–10 years, on the other hand, conservation of weight is admitted by virtue of the same inferences as have been applied to the material, but up to 11 or 12 years conservation of volume is still denied, and by virtue of the converse intuitive reasoning! Moreover, seriations, combinations based on equality, etc., follow exactly the same order of development: at 8 years two quantities of matter that equal a third are equal to each other, but not so two weights (which are, needless to say, independent of perception of volume)! The reason for these separations is naturally to be sought in the intuitive characters of substance, weight and volume, which facilitate or hinder operational combinations. Thus, up to the age of 11 or 12, a particular logical form is still not independent of its concrete content.

FORMAL OPERATIONS

The separations of which we have just seen an example relate to operations affecting similar categories of actions or concepts, even though they apply to distinct fields; since they occur during the same period, they may be called, "horizontal separations". On the other hand, the transition from sensori-motor co-ordinations to representative co-ordinations gives rise, as we have seen, to similar reconstructions involving separations, but since these no longer relate to the same levels they may be called "vertical". Now the building up of formal operations, which begins at about 11 or 12 years, likewise necessitates a complete reconstruction, which serves to transpose "concrete" groupings to a new level of thought, and this reconstruction is character-ized by a series of vertical separations.

Formal thought reaches its fruition during adolescence. The adolescent, unlike the child, is an individual who thinks beyond the present and forms theories about everything, delighting especially in considerations of that which is not. The child, on the other hand, concerns himself only with action in progress and does not form theories, even though an observer notes the periodical recurrence of analogous reactions and may discern a spontaneous systematization in his ideas. This reflective thought, which is characteristic of the adolescent, exists from the age of 11–12 years, from the time, that is, when the subject becomes capable of reasoning in a hypothetico-deductive man-ner, i.e., on the basis of simple assumptions which have no necessary relation to reality or to the subject's beliefs, and from the time when he relies on the necessary validity of an inference (vi formae), as opposed to agreement of the conclusions with experience.

Now, reasoning formally and with mere propositions involves different operations from reasoning about action or reality. Rea-soning that concerns reality consists of a first-degree grouping of operations, so to speak, i.e. internalised actions that have become capable of combination and reversal. Formal thought, on the other hand, consists in reflecting (in the true sense of the word) on these operations and therefore operating on operations or on their results and consequently effecting a second-degree grouping of operations. No doubt the same operational content is involved; the problem is still a matter of classing, serialising, enumerating, measuring, placing or displacing in space or in time, etc. But these classes, series and spatio-temporal relations themselves, as structurings of action and reality, are not what is grouped by formal operations but the propositions that express or "reflect" these operations. Formal operations, therefore, con-sist essentially of "implications" (in the narrow sense of the word) and "contradictions" established between propositions which themselves express classifications, seriations, etc.

We can now see why there is a vertical separation between concrete operations and formal operations, even though the second repeats to some extent the content of the first; the operations in question are indeed not by any means of the same psychological difficulty. Thus, one has only to translate a simple problem of seriation between three terms presented in random order into propositions for this serial addition to become singularly difficult, although, right from the age of 7, it is quite easy as long as it takes the form of a concrete seriation or even of transitive co-ordinations considered in relation to action. The following neat example comes from one of Burt's tests: "Edith is fairer than Susan; Edith is darker than Lily; who is the darkest of the three?" Now this problem is rarely solved before the age of 12. Till then we find reasoning such as the following: Edith and Susan are fair, Edith and Lily are dark, therefore Lily is darkest, Susan is the fairest and Edith in between. In other words, the child of 10 reasons formally as children of 4–5 years do when serialising sticks, and it is not until the age of 12 that he can accomplish with formal problems what he could do with concrete problems of size at the age of 7, and the cause of this is simply that the premises are given as pure verbal postulates and the conclusion is to be drawn *vi formae* without recourse to concrete operations.

We thus see why formal logic and mathematical deduction are still inaccessible to the child and seem to constitute a realm on its own—the realm of "pure" thought which is independent of action. And indeed, whether we are concerned with the particular language—which, like every language, is learned—of mathematical signs (signs which are quite different from symbols in the sense defined above) or with the other system of signs (i.e. the words expressing simple propositions), hypothetico-deductive operations are situated on a different plane from concrete reasoning, since an action affecting signs that are detached from reality is something quite different from an action relating to reality itself or relating to signs attached to this reality. This is why logic dissociates this final stage from the main body of mental development and is in fact limited to axiomatizing characteristic operations instead of replacing them in their living context. This always was its role, but this role certainly gains by being played consciously. Moreover, logic was driven to this course by the very nature of formal operations which, since second-degree operations deal only with signs, are committed to the schematization proper to an axiomatic. But it is the function of the psychology of intelligence to replace the canon of formal operations in its true perspective and to show that it could not have any mental meaning, were it not for the concrete operations that both pave the way for it and provide its content. Formal logic is, according to this view, not an adequate description for the whole of living thought; formal operations constitute solely the structure of the final equilibrium to which concrete operations tend when they are reflected in more general systems linking together the propositions that express them.

THE HIERARCHY OF OPERATIONS AND THEIR PROGRESSIVE DIFFERENTIATION

As we have seen, a response is a functional interaction between subject and objects, and responses may be serialised in an order of genetic succession, based on the increasing distances, spatial and temporal, that characterize the increasingly complex routes followed by these interactions.

Thus, perceptual assimilation and accommodation involve merely a direct and rectilinear form of interaction. Habit has routes that are more complex but shorter, stereotyped and uni-directional. Sensori-motor intelligence introduces reversals and detours; it has access to objects outside the perceptual field and habitual routes and so it goes beyond original distances in space and time but is still limited to the field of the subject's own action. With the beginnings of representative thought and especially with the growth of intuitive thought, intelligence becomes capable of evoking absent objects, and consequently of being applied to invisible realities in the past and partly even in the future. But it still proceeds by way of more or less static figures—half-individual, half-generic images in the case of the preconcept, complex representative configurations, which are still better articulated, in the intuitive period—but they are nevertheless figures, i.e. "stills" of moving reality, which represent only some states or pathways out of the mass of possible routes. Intuitive thought thus provides a map of reality (which sensori-motor intelligence, bound up with immediate reality, could not do), but it is still imaginal, with many blank spaces and without sufficient co-ordinations to pass from one point to another. When groupings of concrete operations appear, these forms are dissolved or blended into the all-embracing plan and decisive progress is made towards the overcoming of distances and the differentiation of routes; thought no longer masters only fixed states or pathways but even deals with changes, so that one can always pass from one point to another and vice versa. Thus, the whole of reality becomes accessible. But it is still only a represented reality; with formal operations there is even more than reality involved, since the world of the possible becomes available for construction and since thought becomes free from the real world. Mathematical creativity is an illustration of this new power.

Now to picture the mechanism of this process of construction and not merely its progressive extension, we must note that each level is characterized by a new co-ordination of the elements provided—already existing in the form of wholes, though of a lower order—by the processes of the previous level.

The sensori-motor schema, the characteristic unit of the

system of pre-symbolic intelligence, thus assimilates perceptual schemata and the schemata relating to learned action (these schemata of perception and habit being of the same lower order, since the first concerns the present state of the object and the second only elementary changes of state). The symbolic schema assimilates sensori-motor schemata with differentiation of function; imitative accommodation is extended into imaginal significants and assimilation determines the significates. The intuitive schema is both a co-ordination and a differentiation of imaginal schemata. The concrete operational schema is a grouping of intuitive schemata, which are promoted, by the very fact of their being grouped, to the rank of reversible operations. Finally, the formal schema is simply a system of second-degree operations, and therefore a grouping operating on concrete groupings.

Each of the transitions from one of these levels to the next is therefore characterized both by a new co-ordination and by a differentiation of the systems constituting the unit of the preceding level. Now these successive differentiations, in their turn, throw light on the undifferentiated nature of the initial mechanisms, and thus we can conceive both of a genealogy of operational groupings as progressive differentiations, and of an explanation of the pre-operational levels as a failure to differentiate the processes involved.

Thus, as we have seen (Chap. 4), sensori-motor intelligence arrives at a kind of empirical grouping of bodily movements, characterized psychologically by actions capable of reversals and detours, and geometrically by what Poincaré called the (experimental) group of displacement. But it goes without saying that, at this elementary level, which precedes all thought, we cannot regard this grouping as an operational system, since it is a system of responses actually effected; the fact is therefore that it is undifferentiated, the displacements in question being at the same time and in every case responses directed towards a goal serving some practical purpose. We might therefore say that at this level spatio-temporal, logico-arithmetical and practical (means and ends) groupings form a global whole and that, in the absence of differentiation, this complex system is incapable of constituting an operational mechanism.

At the end of this period and at the beginning of representative thought, on the other hand, the appearance of the symbol makes possible the first form of differentiation: practical groupings (means and ends) on the one hand, and representation on the other. But this latter is still undifferentiated, logico-arithmetical operations not being distinguished from spatio-temporal operations. In fact, at the intuitive level there are no genuine classes or relations because both are still spatial collections as well as spatio-temporal relationships: hence their intuitive and pre-operational character. At 7–8 years, however, the appearance of operational groupings is characterized precisely by a clear differentiation between logico-arithmetical operations that have become independent (classes, relations and despatialized numbers) and spatio-temporal or infra-logical operations. Lastly, the level of formal operations marks a final differentiation between operations tied to real action and hypothetico-deductive operations concerning pure implications from propositions stated as postulates.

THE DETERMINATION OF "MENTAL AGE"

The knowledge acquired from the psychology of intelligence has given rise to three kinds of applications, which do not, as such, concern our subject, but which yield information for checking theoretical hypotheses.

Everybody knows how Binet, with a view to determining the degree of retardation of the abnormal, came to invent his remarkable metrical scale of intelligence. Binet, a subtle analyst of thought processes, was more aware than anybody of the difficulties of arriving through his measurements at the actual mechanism of intelligence. But precisely because of this feeling of doubt, he had recourse to a kind of psychological probabilism and, in collaboration with Simon, gathered together the most diverse tests and sought to determine frequency of success as a function of age; intelligence is thus assessed by advance or retardation according to the mean statistical age for the correct solutions.

It is indisputable that these tests of mental age have on the whole lived up to what was expected of them: a rapid and convenient estimation of an individual's general level. But it is no less obvious that they simply measure a "yield" without reaching constructive operations themselves. As Piéron quite rightly pointed out, intelligence conceived in these terms is essentially a value-judgment applied to complex behaviour.

On the other hand, tests have multiplied apace and attempts have been made to distinguish them according to the different special aptitudes they measure. In the field of intelligence itself, tests of reasoning, comprehension, knowledge etc., have thus been devised. So the problem is to work out the correlations between these statistical results, in the hope of distinguishing and measuring the various factors involved in the inner mechanism of thought. Spearman and his school, in particular, have applied themselves to this task, using precise statistical methods[1] and they have arrived at the hypothesis that certain constant factors are involved. The most general of these Spearman called "the 'g' factor" and its value is related to the individual's intelligence. But, as this writer himself insisted, the "g" factor is simply expressed as "general intelligence", i.e. the degree of efficiency common to all the subject's aptitudes or, one might almost say, the quality of neural and psychological organisation making a mental task easier for one individual than for others.

Finally, there has been an attempt to react in another way

[1] Calculation of "tetrad differences" or correlations between correlations.

against the empiricism of simple measures of yield, namely, by trying to ascertain the actual operations that a given individual has at his disposal; the term "operation" is here taken in a limited sense as relative to genetic construction, as we have treated it in this work. In this way, B. Inhelder has made use of the concept of a "grouping" in testing reasoning-power. She was able to show that the order of acquiring concepts of conservation of substance, weight and volume recurs in its entirety in mental deficients; the last of these three constants (present only in slightly backward individuals and unknown in really deficient cases) is never found without the other two, nor the second without the first, while conservation of substance occurs without conservation of weight and volume and that of substance and weight without that of volume. She was able to distinguish moronism from imbecility by the presence of concrete groupings (of which the imbecile is not capable), and slight backwardness by an inability to reason formally, i.e. by incompleteness of operational construction.[1] This is one of the first applications of a method which could be developed further for determining levels of intelligence in general.

---

[1] B. Inhelder, Le diagnostic du raisonnement chez les débiles mentaux, Delachaux et Niestlé, 1944.

6

SOCIAL FACTORS IN INTELLECTUAL DEVELOPMENT

The human being is immersed right from birth in a social environment which affects him just as much as his physical environment. Society, even more, in a sense, than the physical environment, changes the very structure of the individual, because it not only compels him to recognize facts, but also provides him with a ready-made system of signs, which modify his thought; it presents him with new values and it imposes on him an infinite series of obligations. It is therefore quite evident that social life affects intelligence through the three media of language (signs), the content of interaction (intellectual values) and rules imposed on thought (collective logical or pre-logical norms).

Certainly, it is necessary for sociology to envisage society as a whole, even though this whole, which is quite distinct from the sum of the individuals composing it, is only the totality of relations or interaction between these individuals. Every relation between individuals (from two onwards) literally modifies them and therefore immediately constitutes a whole, so that the whole formed by society is not so much a thing, a being or a cause as a system of relations. But these relations are extremely numerous and complex, since, in fact, they constitute just as much a continuous plot in history, through the action of successive generations on each other, as a synchronous system of equilibrium at each moment of history. It is therefore legitimate to adopt statistical language and to speak of "society" as a coherent whole (in the same way as a *Gestalt* is the resultant of a statistical system of relations). But it is essential to remember the statistical nature of statements in sociological language, since to forget this would be to attribute a mythological sense to the words. In the sociology of thought it might even be asked whether it would not be better to replace the usual global language by an enumeration of the types of relation involved (types which, needless to say, are likewise statistical).

When we are concerned with psychology, on the other hand, i.e. when the unit of reference is the individual modified by social relations, rather than the complex or complexes of relations as such, it becomes quite wrong to content oneself with statistical terms, since these are too general. The "effect of social life" is a concept which is just as vague as that of "the effect of the physical environment" if it is not described in detail. From birth to adult life, the human being is subject, as nobody denies, to social pressures, but these pressures are of extremely varied types and are subject to a certain order of development. Just as the physical environment is not imposed on developing intelligence all at once or as a single entity, but in such a way that acquisitions can be followed step by step as a function of experience, and especially as a function of the kinds of assimilation or accommodation—varying greatly according to mental level—that govern these acquisitions, so the social environment gives rise to interactions between the developing individual and his fellow, interactions that differ greatly from one another and succeed one another according to definite laws. These types of interaction and these laws of succession are what the psychologist must carefully establish, lest he simplify the task to the extent of giving it up in favour of the problems of sociology. Now there is no longer any reason for conflict between this science and psychology once one recognises the extent to which the structure of the individual is modified by these interactions; both of these two disciplines, therefore, stand to gain by an investigation that goes beyond a global analysis and undertakes to analyse relations.

THE SOCIALIZATION OF INDIVIDUAL INTELLIGENCE

The interaction with his social environment in which the individual indulges varies widely in nature according to his level of development, and consequently in its turn it modifies the individual's mental structure in an equally varied manner.

During the sensori-motor period the infant is, of course, already subject to manifold social influences; people afford him the greatest pleasures known to his limited experience—from food to the warmth of the affection which surrounds him—people gather round him, smile at him, amuse him, calm him; they inculcate habits and regular courses of conduct linked to signals and words; some behaviour is already forbidden and he is scolded. In short, seen from without, the infant is in the midst of a multitude of relations which forerun the signs, values and rules of subsequent social life. But from the point of view of the subject himself, the social environment is still not essentially distinct from the physical environment, at least up to the fifth of the stages of sensori-motor intelligence that we have distinguished (Chap. 4). The signs that are used to affect him are, as far as he is concerned, only indices or signals. The rules imposed on him are not yet obligations of conscience and he confuses them with the regularity characteristic of habit. As for people, they are seen as pictures like all the pictures which constitute reality, but they are particularly active, unpredictable and the source of the most intense feelings. The infant reacts to them in the same way as to objects, namely with gestures that happen to cause them to continue interesting actions, and with various cries, but there is still as yet no interchange of thought, since at this level the child does not know thought; nor, consequently, is there any profound modification of intellectual structures by the social life surrounding him.[1]

With the acquisition of language, however, i.e. with the advent of the symbolic and intuitive periods, new social relations appear which enrich and transform the individual's thought. But in this context three points should be noted.

In the first place, the system of collective signs does not create the symbolic function, but naturally develops it to a degree that the individual by himself would never know. Nevertheless, the sign as such, conventional (arbitrary) and ready-made, is not an adequate medium of expression for the young child's thought; he is not satisfied with speaking, he must needs "play out" what he thinks and symbolize his ideas by means of gestures or objects, and represent things by imitation, drawing and construction. In short, from the point of view of expression itself, the child at the outset is still midway between the use of the collective sign and that of the individual symbol, both still being necessary, no doubt, but the second being much more so in the child than in the adult.

In the second place, language conveys to the individual an already prepared system of ideas, classifications, relations—in short, an inexhaustible stock of concepts which are

---

[1] From the affective point of view, it is no doubt only at the stage at which the notion of an object is formed that there is a projection of affectivity on to people conceived as similar centres of independent action.

reconstructed in each individual after the age-old pattern which previously moulded earlier generations. But it goes without saying that the child begins by borrowing from this collection only as much as suits him, remaining disdainfully ignorant of everything that exceeds his mental level. And again, that which is borrowed is assimilated in accordance with his intellectual structure; a word intended to carry a general concept at first engenders only a half-individual, half-socialised pre-concept (the word "bird" thus evokes the familiar canary, etc.).

There remain, in the third place, the actual relations that the subject maintains with his fellow beings, i.e. "synchronous" relations, as opposed to the "diachronic" processes that influence the child's acquisition of language and the modes of thought that are associated with it. Now these synchronous relations are at first essential; when conversing with his family, the child will at every moment see his thoughts approved or contradicted, and he will discover a vast world of thought external to himself, which will instruct or impress him in various ways. From the point of view of intelligence (which is all that concerns us here), he will therefore be led to an ever more intensive exchange of intellectual values and will be forced to accept an ever-increasing number of obligatory truths (ready-made ideas and true norms of reasoning).

But here again we must not exaggerate or confuse capacities for assimilation as they appear in intuitive thought with the form they take at the operational level. In fact, as we have seen in connection with the adaptation of thought to the physical environment, intuitive thought, which is dominant up to the end of early childhood (7 years), is characterized by a disequilibrium, still unresolved, between assimilation and accommodation. An intuitive relation always results from a "centring" of thought depending on one's own action, as opposed to a "grouping" of all the relations involved; thus the equivalence between two series of objects is recognised only in relation to the act of making them correspond, and is lost as soon as this action is replaced by another. Intuitive thought, therefore, always evinces a distorting egocentricity, since the relation that is recognised is related to the subject's action and not decentralised into an objective system.[1]

Conversely, and precisely because intuitive thought is from moment to moment "centred" on a given relation, it is phenomenalistic and grasps only the perceptual appearance of reality. It is therefore a prey to suggestion coming from immediate experience, which it copies and imitates instead of correcting. Now the reaction of intelligence at this level to the social environment is exactly parallel to its reaction to the physical environment, and this is self-evident, since the two kinds of experience are indistinguishable in reality.

For one thing, however dependent he may be on surrounding intellectual influences, the young child assimilates them in his own way. He reduces them to his point of view and therefore distorts them without realizing it, simply because he cannot yet distinguish his point of view from that of others through failure to co-ordinate or "group" the points of view. Thus, both on the social and on the physical plane, he is egocentric through ignorance of his own subjectivity. For example, he can show his right hand but confuses the right-left relationship in a partner facing him, since he cannot see another point of view, either socially or geometrically; similarly, we have noted how, in problems of perspective, he first attributes his own view of things to others; in questions involving time there are even cases where a young child, while stating that his father is much older than himself, believes him to have been born "after" himself, since he cannot "remember" what he did before! In short, intuitive centralisa-

---

[1] Wallon, who has criticised the concept of egocentricity, nevertheless retains the phenomenon itself, as he neatly shows when he says that the young child thinks in the optative and not in the indicative mood.

tion, as opposed to operational decentralisation, is thus reinforced by an unconscious and therefore all the more systematic primacy of his own point of view. This intellectual egocentricity is in both cases nothing more than a lack of co-ordination, a failure to "group" relations with other individuals as well as with other objects. There is nothing here that is not perfectly natural; the primacy of one's own point of view, like intuitive centralisation in accordance with the subject's own action, is merely the expression of an original failure to differentiate, of an assimilation that distorts because it is determined by the only point of view that is possible at first. Actually, such a failure to differentiate is inevitable, since the distinction between different points of view, as well as their co-ordination, requires the activity of intelligence.

But, because the initial egocentricity results from a simple lack of differentiation between *ego* and *alter*, the subject finds himself exposed during the very same period to all the suggestions and constraints of his fellows, and he accommodates himself without question, simply because he is not conscious of the private nature of his viewpoint (it thus frequently happens that young children do not realize that they are imitating, and believe that they have originated the behaviour in question, just as they may attribute their own private ideas to others). That is why the period of maximum egocentricity in the course of development coincides with the maximum pressure from the examples and opinions of his fellows, and the combination of assimilation to the self and accommodation to surrounding models is just as explicable as that of the egocentricity and phenomenalism characterizing the first intuition of physical relations.

However, it is obvious that under these conditions (all of which involve the absence of "grouping") the coercions of other people would not be enough to engender a logic in the child's mind, even if the truths that they imposed were rational in content; repeating correct ideas, even if one believes that they originate from oneself, is not the same as reasoning correctly. On the contrary, in order to teach others to reason logically it is indispensable that there should be established between them and oneself those simultaneous relationships of differentiation and reciprocity which characterize the co-ordination of viewpoints.

In short, at the pre-operational levels, extending from the appearance of language to the age of about 7–8 years, the structures associated with the beginnings of thought preclude the formation of the co-operative social functions which are indispensable for logic to be formed. Oscillating between distorting egocentricity and passive acceptance of intellectual suggestion, the child is, therefore, not yet subject to a socialization of intelligence which could profoundly modify its mechanism.

At the stage at which groupings of concrete operations and particularly when those of formal operations are constructed, on the other hand, the problem of the respective roles of social interaction and individual structures in the development of thought arises in all its acuteness. Genuine logic, which is formed during these two periods, shows in fact social characteristics of two kinds, and we have to decide whether these result from the appearance of groupings or whether they are the cause of them. On the one hand, the more intuitions articulate themselves and end by grouping themselves operationally, the more adept the child becomes at co-operation, a social relationship which is quite distinct from coercion in that it involves a reciprocity between individuals who know how to differentiate their viewpoints. As far as intelligence is concerned, co-operation is thus an objectively conducted discussion (out of which arises internalized discussion, i.e. deliberation or reflection), collaboration in work, exchange of ideas, mutual control (the origin of the need for verification and demonstration), etc. It is therefore clear that co-operation is the first of a series of forms of behaviour which are important for the constitution and development of logic. On the other hand, from the psychological point of view—which is our point of view here—logic itself does not consist solely of a system of free operations; it expresses itself as a complex of states of awareness, intellectual feelings and responses, all of which are characterized by certain obligations whose social character is difficult to deny, be it primary or derived. Considered from this angle, logic requires common rules or norms; it is a morality of thinking imposed and sanctioned by others. Thus, the obligation not to contradict oneself is not simply a conditional necessity (a "hypothetical imperative") for anybody who accepts the exigencies of operational activity; it is also a moral "categorical" imperative, inasmuch as it is indispensable for intellectual interaction and co-operation. And, indeed, the child first seeks to avoid contradicting himself when he is in the presence of others. In the same way, objectively, the need for verification, the need for words and ideas to keep their meaning constant, etc. are as much social obligations as conditions of operational thought.

One question now arises which is inescapable: is the "grouping" the cause or the effect of co-operation? Grouping is a co-ordination of operations, i.e. of actions accessible to the individual. Co-operation is a co-ordination of viewpoints or of actions emanating from different individuals. Their affinity is thus obvious, but does operational development within the individual enable him to co-operate with others, or does external co-operation, later internalized in the individual, compel him to group his actions in operational systems?

OPERATIONAL "GROUPINGS" AND CO-OPERATION

To such a question there must of course be two distinct and complementary answers. One is that without interchange of thought and co-operation with others the individual would never come to group his operations into a coherent whole: in this sense, therefore, operational grouping presupposes social life. But, on the other hand, actual exchanges of thought obey a law of equilibrium which again could only be an operational grouping, since to co-operate is also to co-ordinate operations. The grouping is therefore a form of equilibrium of inter-individual actions as well as of individual actions, and it thus regains its autonomy at the very core of social life.

It is in fact very difficult to understand how the individual would come to group his operations in any precise manner, and consequently to change his intuitive representations into transitive, reversible, identical and associative operations, without interchange of thought. The grouping consists essentially in a freeing of the individual's perceptions and spontaneous intuitions from the egocentric viewpoint, in order to construct a system of relations such that one can pass from one term or relation to another belonging to any viewpoint. The grouping is therefore by its very nature a co-ordination of viewpoints and, in effect, that means a co-ordination between observers, and therefore a form of co-operation between several individuals.

Let us suppose, however, with common sense, that a superior individual, by ceaselessly shifting his viewpoints, manages all by himself to co-ordinate them with one another so that their grouping is assured. But how could a single individual, even if he were endowed with sufficient experience, manage to recall his previous viewpoints, i.e. all the relations he has perceived at one time or another but which he no longer perceives? If he were capable of this, he must have succeeded in establishing a kind of interaction between his various successive states, i.e. he has built up, by continual agreement with himself, a system of notation which could consolidate his memories and translate them into a representative language; he would then have achieved a "society" between his different "selves"! In fact, it is precisely by a constant interchange of thought with others that we are able to decentralise ourselves in this way, to co-ordinate internally relations deriving from different viewpoints. In particular, it is very difficult to see how concepts could conserve their permanent meanings and their definitions were it not for co-operation; the very reversibility of thought is thus bound up with a collective conservation without which individual thought would have only an infinitely more restricted mobility at its disposal.

But, granting all this and admitting that logical thought is necessarily social, the fact remains that the laws of grouping constitute general forms of equilibrium which express both the equilibrium of inter-individual interaction and that of the operations of which every socialized individual is capable when he reasons internally in terms of his most personal and original ideas. To say that an individual arrives at logic only through co-operation thus simply amounts to asserting that the equilibrium of his operations is dependent on an infinite capacity for interaction with other people and therefore on a complete reciprocity. But this statement contains nothing that is not obvious, since the grouping within him is already nothing more nor less than a system of reciprocities.

Moreover, if we enquire what an interaction of thought between individuals is, we find that it consists essentially of systems of correspondences, and therefore of well-defined "groupings"; to a certain relation established from A's viewpoint there corresponds, after interaction, such and such a relation from B's viewpoint, and a certain operation executed by A corresponds (whether it be equivalent or merely reciprocal) to a certain operation executed by B. These correspondences are what, for each proposition stated by A or B, determine the agreement (or, in the case of non-correspondence, the disagreement) of the parties, their obligation to conserve admitted propositions and the lasting validity of the latter in the course of subsequent interchanges. Intellectual interaction between individuals is thus comparable to a vast game of chess, which is carried on unremittingly and in such a way that each action carried out with respect to a particular item involves a series of equivalent or complementary actions on the part of the opponent; laws of grouping are nothing more or less than the various rules ensuring the reciprocity of the players and the consistency of their play.

More precisely, every grouping within individuals is a system of operations, and co-operation constitutes the system of operations executed in common, i.e. co-operations, in the true sense of the word.

It would be incorrect, however, to conclude that the laws of grouping are superior both to co-operation and to individual thought; they only form, we repeat, laws of equilibrium, and express merely the particular form of equilibrium that is reached, on the one hand, when society no longer exerts distorting constraints on the individual but inspires and maintains the free play of his mental processes and, on the other hand, when this free play of thought in each individual no longer distorts that of other people and no longer distorts objects, but has regard for the reciprocity between different activities. Defined in this way, this form of equilibrium could not be considered either as a result of individual thought alone or as an exclusively social product; internal operational activity and external co-operation are merely, taking these words in their most precise senses, two complementary aspects of one and the same whole, since the equilibrium of the one depends on that of the other. Moreover, since an equilibrium is never completely achieved in practice, the ideal form which it would ultimately assume has to be imagined, and it is this ideal equilibrium that is described axiomatically by logic. The logician, therefore, works with the ideal (as opposed to the real) and is entitled to confine himself to this, since the equilibrium with which he deals is never fully achieved, and since it is constantly projected still higher, as new actual constructions appear. As for the sociologist and the psychologist, they can only consult each other to ascertain how this equilibration is realized in practice.

CONCLUSION

RHYTHMS, REGULATIONS AND GROUPINGS

Intelligence, viewed as a whole, takes the form of a structuring which impresses certain patterns on the interaction between the subject or subjects and near or distant surrounding objects. Its originality resides essentially in the nature of the patterns that it constructs to this effect.

Life itself is a "creator of patterns", as Brachet has remarked.[1] Certainly, these biological "patterns" are those of the organism, of each of its organs and of the physical interaction with the environment which they safeguard. But in instinct, anatomico-physiological patterns are paralleled by functional interactions, i.e. by "patterns" of behaviour. In fact, instinct is only a functional extension of the structure of organs; the beak of a woodpecker finds its extension in the pecking instinct, a digging paw

---

[1] And, from this point of view, the assimilatory schemata which control the development of intelligence are comparable to the "organizers" which intervene in embryological development.

in the burrowing instinct, etc. Instinct is the logic of organs, and that is how it arrives at responses which, if they were realized at the level of genuine operations, would in many cases imply a prodigious intelligence, although its "patterns" may at first sight seem analogous (as in seeking for an object outside the perceptual field and at various distances).

Habit and perception constitute other "patterns", as Gestalt theory has insisted, working out the laws of their organization. Intuitive thought reveals still others. As for operational intelligence, this, as we have repeatedly seen, is characterized by mobile and reversible "patterns" which are constituted by groups or groupings.

If we wish to bring what we have learned from an analysis of the operations of intelligence into line with the biological considerations with which we started (Chap. I), we have to end by seeing operational structures in their relation to the mass of possible "patterns". Now, an operational act may, in its content, closely resemble an intuitive act, a sensori-motor or perceptual act and even an instinctive act; a geometrical figure may thus be the product of a logical construction, a pre-operational intuition, a perception, an automatic habit and even a building instinct. The difference between the various levels does not, therefore, depend on the content, i.e. on a "pattern" somehow materialized, which results from the act,[1] but on the "pattern" of the act itself and of its progressive organization. In the case of reflective thought which has attained an equilibrium, this pattern consists of a certain "grouping" of operations. In the continuum of cases between perception and intuitive thought, the pattern of the response is that of an adjustment occurring at various speeds (sometimes almost instantaneously), but always

---

[1] It is to be noted that this external pattern is precisely what the Gestalt theory has especially insisted on, which was bound to induce an undue neglect of genetic construction.

functioning by "regulations". In the case of instinctive or reflex behaviour, we are confronted with a framework which is relatively complete, rigid, and self-contained and which functions by periodic repetitions or "rhythms". The order of succession of the fundamental structures or "patterns" concerned in the development of intelligence would thus be: rhythms, regulations, groupings.

The organic or instinctive needs which motivate elementary behaviour are in fact periodic and therefore follow a rhythmic structure: hunger, thirst, sexual appetite, etc. As regards the reflex frameworks which allow of their satisfaction and constitute the underlying structure of mental life, we now know well enough that they form complex systems and do not result from an additive combination of elementary reactions; the locomotion of a biped and, even more so, of a quadruped (the organization of which, according to Graham Brown, evinces an overall rhythm which dominates and even precedes differentiated reflexes), the exceedingly complex reflexes which govern sucking in the neonate, etc. and even the impulsive movements which characterize the infant's behaviour, show a way of functioning whose rhythmical form is obvious. The instinctive behaviour, often highly specialized, of animals also consists of a well defined chain of responses, which take the form of a definite rhythm, since they are repeated periodically at constant intervals. Rhythm, therefore, characterizes the functions that are at the junction between organic and mental life, and this is so universally true that even in the field of elementary perception or sensation the measurement of sensitivity reveals the existence of primitive rhythms which completely elude the subject's awareness; rhythm is likewise at the root of all effector functions including those that constitute motor habit.

Now, rhythm shows a structure which must be borne in mind if we are to see intelligence in its relation to the mass of living "patterns", for it involves a way of linking elements together which already heralds in an elementary form what will appear as the reversibility characteristic of the higher mental processes. Whether we are concerned with particular reflex facilitations and inhibitions or, more generally, with a succession of responses in alternating and opposite directions, the rhythm schema always involves, in one way or another, the alternation of two antagonistic processes, the one functioning in the direction A–B and the other in the opposite direction B–A. It is true that in a system of perceptual regulations, whether intuitive or relating to responses co-ordinated according to experience, there also exist processes which are orientated in opposite directions; but they follow each other irregularly and in relation to "displacements of equilibrium" occasioned by a new external situation. The antagonistic responses of rhythm, on the other hand, are governed by the actual internal (and hereditary) framework, and consequently manifest a regularity which is much more rigid and self-sufficient. There is an even greater difference between rhythm and the "converse operations" characterizing intelligent reversibility, which are intentional and associated with the infinitely mobile combinations of the "grouping".

Hereditary rhythm thus ensures a certain conservation of responses which in no way precludes their being complex or comparatively flexible (the rigidity of instincts has been exaggerated). But, in so far as one is confined to innate mechanisms, this conservation of periodic schemata evinces a systematic lack of differentiation between the assimilation of objects to the subject's activity and the accommodation of the latter to possible changes in the external situation.

In the case of learning by experience, however, accommodation is differentiated and, as this process progresses, elementary rhythms are integrated into vaster systems which no longer show any regular periodicity. On the other hand, a second fundamental structure now appears which continues the work of the original periodicity and consists of regulations;[1] these we have encountered from perception right up to pre-operational intuitions. A perception, for example, always constitutes a complex system of relations and may thus be considered as the momentary form of equilibrium reached by a multitude of elemental sensory rhythms which combine or conflict in various ways. This system tends to be conserved as a totality as long as external phenomena remain unchanged, but, once they are modified, accommodation to new phenomena involves a "displacement of equilibrium". But these displacements are not uncontrolled and the equilibrium that is re-established by assimilation to previous perceptual schemata shows a tendency to react in the opposite direction to that of the external change.[2] There is therefore regulation, i.e. the occurrence of antagonistic processes comparable to those already manifest in periodic responses, but here the phenomenon occurs on a larger scale, which is much more complex and far-reaching and does not necessarily show periodicity.

The structure characterized by the existence of regulations is not peculiar to perception. It occurs also in the "corrections" belonging to motor learning. The whole of sensori-motor development in general, up to and including the various levels of sensori-motor intelligence, reveals analogous systems. Only in one special case, namely that of true displacements with reversals and detours, does the system tend to reach reversibility and so herald the grouping, but with the restrictions that we have seen. In most cases, on the other hand, a regulation, while moderating and correcting disturbing modifications and therefore working in the opposite direction to earlier changes, does not attain

---

[1] We refer here, of course, to structural regulations and not to the dynamic regulations which, according to Janet, etc., characterize affective life at these same levels.

[2] E.g. see Delbœuf's illusion quoted on p. 67.

complete reversibility for lack of a complete adjustment between assimilation and accommodation.

When thought begins to appear, intuitive centralisations and the egocentricity of successively constructed relations restrict thought to its irreversible state, as has been seen (Chap. 5) in connection with non-conservation. Intuitive changes, therefore, are only "compensated" by a system of regulations which, in the course of the internal trial-and-error of representation, gradually harmonize mental assimilation and accommodation and monopolize the control of non-operational thought.

Now it is easy to see that these regulations themselves, whose various types extend from elementary habits and perceptions to the threshold of operations, grow out of the original "rhythms" without any real discontinuity. We should, first of all, remember that the first acquisitions to follow the exercise of hereditary connections also present a form of rhythm; the "circular reactions", which are the first actively acquired habits, consist of repetitions with a clearly visible periodicity. Perceptual estimations of sizes or complex shapes (and not only those of absolute intensity) again reveal the existence of a continuous oscillation about a definite point of equilibrium. Similarly, it may be assumed that components analogous to those determining the alternating and antagonistic phases of rhythm (A–B and B–A) also occur in a complex system subject to regulations, but they then appear simultaneously and in momentary equilibrium with each other, instead of each alternately coming to the fore; that is why, when this equilibrium is changed, there is a "displacement of equilibrium" and the appearance of a tendency to resist external modifications, i.e. to moderate the change which is undergone (as physicists say in connection with the well-known mechanism described by Le Châtelier). It may therefore be understood that when components of action constitute complex static systems, responses orientated in opposite directions (whose alternation formerly brought about the distinct and successive phases of rhythm) are synchronized and represent the elements of the system's equilibrium. In the event of external changes, the equilibrium is upset through the accentuation of one of the tendencies involved, but this accentuation is sooner or later checked by the intervention of the opposite tendency; this reversal of direction is then what is meant by regulation.

We now understand the nature of the reversibility characteristic of operational intelligence, and the way in which the converse operations of *grouping* derive from regulations, and not only intuitive but even sensori-motor and perceptual regulations. Reflex rhythms are not reversible as wholes but are orientated in a definite direction; execution of a movement (or a complex of movements), the termination and a return to the point of origin in order to repeat it in the same direction such are its successive phases, and if the return (or antagonistic) phase reverses the original movements, this is not a case of a second action having the same value as the positive phase, but a retraction leading to a new beginning in the same direction. Nevertheless, the antagonistic phase of rhythm marks the beginnings of regulations and, beyond this, of the "converse operations" of intelligence, and so all rhythm can be regarded as a system of alternating regulations combined into a single unit of successive elements. As for regulation, which would thus constitute the product of a complex rhythm whose components have become simultaneous, this characterizes behaviour which is still irreversible but whose reversibility is an advance on previous behaviour. Even at the perceptual level, the reversal of an illusion implies that a relation (e.g. of similarity) outweighs the opposite relation (difference) after a certain degree of exaggeration of the latter, and vice versa. In the field of intuitive thought this is even clearer; the relation neglected by the centring of attention when it concentrates on another relation dominates the latter in its turn when the error exceeds certain limits. Decentralisation, which is the source of regulation, leads in this case to an intuitive equivalent of converse operations, especially when anticipations and representative reconstructions increase its range and make it almost instantaneous, which occurs more and more when the level of "articulated intuitions" is reached (Chap. 5). Regulation has thus only to achieve complete compensations (towards which, in fact, articulated intuitions tend) for the operation to appear by this very fact; operations are, indeed, merely a system of co-ordinated changes which have become reversible regardless of how they are built up.

So, in the most concrete and precise sense, it is possible to regard the operational groupings of intelligence as the final "pattern" of equilibrium towards which sensori-motor and representative functions tend in the course of their development, and this conception enables us to understand the fundamental functional unity of mental growth, while at the same time we may note the essential differences between the structures characterizing successive levels. Once complete reversibility has been attained—which is the limit of a continuous process, but a limit with quite different properties from those of previous phases, since it marks the advent of equilibrium—the aggregates which were hitherto rigid have become capable of a flexibility of composition which secures their stability since then, whatever operations are executed, accommodation to experience is in permanent equilibrium with assimilation, which is promoted by this very fact to the rank of a necessary deduction.

Rhythm, regulations and "grouping" thus constitute the three phases of the developmental mechanism which connects intelligence with the morphogenetic potentialities of life itself, and enables it to realize adaptations which are both unlimited and mutually equilibrated, adaptations which are impossible to realize at the organic level.

SHORT BIBLIOGRAPHY

CHAPTER 1

Bühler, K. *Die Krise der Psychologie*, Jena (Fischer), 2nd ed., 1929.

Claparède, Ed., "La Psychologie de l'intelligence", *Scientia* (1917), vol. 22, pp. 253–268.

Köhler, W., *Gestalt Psychology*, London, 1929.

Lewin, K., *Principles of Topological Psychology*, London, (McGraw-Hill), 1935.

Montpellier, G. de, *Conduites intelligentes et psychisme chez l'animal et chez l'homme*, Louvain and Paris (Vrin), 1946.

CHAPTER 2

Binet, A., *Etude expérimentale de l'Intelligence*, Paris (Schleicher), 1903.

Burloud, A., *La Pensée d'après les recherches expérimentales de Watt, de Messer et de Bühler*, Paris (Alcan), 1927 (includes references for these three writers).

Delacroix, H., "La Psychologie de la raison" (in) *Traité de Psychologie* by Dumas, 2nd ed. vol. 1, pp. 198–305 Paris (Alcan), 1936.

Lindworsky, I., *Das Schlussfolgernde Denken*, Freiburg-im-Breisgau, 1916.

Piaget, J., *Classes, relations et nombres. Essai sur les "Groupements" de la logistique et la réversibilité de la pensée*, Paris (Vrin), 1942.

Selz, O., *Zur Psychologie des produktiven Denkens und des Irrtums*, Bonn, 1924.

CHAPTER 3

Duncker, K., *Zur Psychologie des produktiven Denkens*, Berlin, 1935.
Guillaume, P., *La Psychologie de la forme*, Paris (Flammarion), 1936.
Köhler, W., *The Mentality of Apes*, London, 1924.
Piaget, J., and Lambercier, M., "Recherches sur le développement des perceptions," *I* to *VIII*, *Archives de Psychologie*, Geneva, 1943–1946.
Wertheimer, M., *Über Schlussprozesse im produktiven Denken*, Berlin, 1920.

CHAPTER 4

Claparède, Ed., "La Genèse de l'hypothèse", *Archives de Psychologie* (Geneva), 1934.
Guillaume, P., *La Formation des habitudes*, Paris (Alcan), 1936.
Hull, C. L., *Principles of Behavior*, New York, 1943.
Krechevsky, I., "The Docile Nature of Hypotheses", *J. Comp. Psychol.*, 1933, vol. 15, pp. 425–443.
Piaget, J., *La Naissance de l'intelligence chez l'enfant* Neuchatel (Delachaux et Niestlé), 1936.
Piaget, J., *La Construction du réel chez l'enfant*, ibid., 1937.
Spearman, C., *The Nature of Intelligence*, London, 1923.
Thorndike, E. L., *The Fundamentals of Learning*, New York (Teacher's College), 1932.
Tolman, E. C., "A Behavioristic Theory of Ideas", *Psychol. Rev.*, vol. 33, pp. 352–369, 1926.

CHAPTERS 5 AND 6

Bühler, C., *From Birth to Maturity*, London, 1935.
Bühler, C., *Mental Development of the Child*, London (Kegan Paul), 1933.
Inhelder, B., *Le Diagnostic du raisonnement chez les débiles mentaux*, Neuchatel (Delachaux et Niestlé), 1944.
Janet, P., *L'Intelligence avant le langage*, Paris (Flammarion), 1935.
Janet, P., *Les Débuts de l'intelligence*, ibid. 1936.

Piaget, J., *La Formation du symbole chez l'enfant*, Neuchatel (Delachaux and Niestlé), 1945.

Piaget, J., *Les Notions de mouvement et de vitesse chez l'enfant*, Paris (Univ. Press), 1946.

Piaget, J., and Szeminska, A., *La Genèse du nombre chez l'enfant*, Neuchatel (Delachaux et Niestlé), 1941.

Piaget, J. and Inhelder, B., *Le Développement des quantités chez l'enfant*, ibid. 1941.

Rey, A., *L'Intelligence pratique chez l'enfant*, Paris (Flammarion), 1942.

Rey, A., *L'Origine de la Pensée chez l'enfant*, Paris (Univ. Press), 1945.

INDEX OF SUBJECTS

INDEX OF NAMES

Routledge Classics – order more now!
Available from all good bookshops.
Credit card orders can be made on our Customer Hotlines:
UK/RoW: +44 (0)8700 768853   US/Canada: (1) 800 634 7064
You can also buy online at: www.routledge.com

TITLE	AUTHOR	ISBN	PRICES UK / US / CAN		
COMPLETE SET - 30 Volumes	Various	0-415-25410-8	**£235** (rrp: £266.70)		
PSYCHOLOGY					
Écrits	Jacques Lacan	0-415-25392-6	£9.99	n/a	n/a
Leonardo da Vinci	Sigmund Freud	0-415-25386-1	£6.99	n/a	n/a
Modern Man in Search of a Soul	Carl Gustav Jung	0-415-25390-X	£7.99	n/a	n/a
On the Nature of the Psyche	Carl Gustav Jung	0-415-25391-8	£7.99	n/a	n/a
The Psychology of Intelligence	Jean Piaget	0-415-25401-9	£9.99	$14.95	$22.95
Totem and Taboo	Sigmund Freud	0-415-25387-X	£7.99	n/a	n/a
LITERATURE					
Principles of Literary Criticism	I. A. Richards	0-415-25402-7	£9.99	$14.95	$22.95
Shakespeare's Bawdy	Eric Partridge	0-415-25400-0	£7.99	$12.95	$19.95
The Pursuit of Signs	Jonathan Culler	0-415-25382-9	£9.99	n/a	n/a
The Wheel of Fire	G. Wilson Knight	0-415-25395-0	£9.99	$14.95	$22.95
What is Literature?	Jean-Paul Sartre	0-415-25404-3	£7.99	n/a	n/a
PHILOSOPHY					
Madness and Civilization	Michel Foucault	0-415-25385-3	£7.99	n/a	n/a
The Fear of Freedom	Erich Fromm	0-415-25388-8	£9.99	n/a	n/a
The Sovereignty of Good	Iris Murdoch	0-415-25399-3	£6.99	$11.95	$17.95
Wickedness	Mary Midgley	0-415-25398-5	£7.99	$12.95	$19.95
Writing and Difference	Jacques Derrida	0-415-25383-7	£9.99	n/a	n/a
Tractatus Logico-Philosophicus	Ludwig Wittgenstein	0-415-25408-6	£7.99	$14.95	$22.95
RELIGION AND ANTHROPOLOGY					
Sex and Repression in Savage Society	Bronislaw Malinowski	0-415-25554-6	£9.99	n/a	n/a
Oppression and Liberty	Simone Weil	0-415-25407-8	£7.99	n/a	$19.95
The Occult Philosophy in the Elizabethan Age	Frances Yates	0-415-25409-4	£9.99	$14.95	$22.95
A General Theory of Magic	Marcel Mauss	0-415-25396-9	£7.99	$12.95	$19.95
Medicine, Magic and Religion	W. H. R. Rivers	0-415-25403-5	£7.99	$12.95	$19.95
Myth and Meaning	Claude Lévi-Strauss	0-415-25394-2	£6.99	n/a	n/a
HISTORY					
The Course of German History	A. J. P. Taylor	0-415-25405-1	£7.99	$12.95	$19.95
The French Revolution	Georges Lefebvre	0-415-25393-4	£9.99	n/a	n/a
SOCIAL AND CULTURAL THEORY					
The Culture Industry	Theodor Adorno	0-415-25380-2	£9.99	$14.95	$22.95
The Protestant Ethic and the Spirit of Capitalism	Max Weber	0-415-25406-X	£9.99	$15.95	$23.95
The Road to Serfdom	F. A. Hayek	0-415-25389-6	£9.99	n/a	n/a
Understanding Media	Marshall McLuhan	0-415-25397-7	£9.99	n/a	n/a
SCIENCE					
Relativity	Albert Einstein	0-415-25384-5	£7.99	$12.95	$19.95

Contact our Customer Hotlines for details of postage and packing charges where applicable.
All prices are subject to change without notification.

Get inside a great mind.